Project AIR FORCE

P9-DTT-336

THE CHANGING ROLE OF THE U.S. MILITARY IN SPACE

Daniel Gonzales

Prepared for the United States Air Force

RAND

The research reported here was sponsored by the United States Air Force under Contract F49642-96-C-0001. Further information may be obtained from the Strategic Planning Division, Directorate of Plans, Hq USAF.

Library of Congress Cataloging-in-Publication Data

Gonzales, Daniel, 1956–   .
    The changing role of the U.S. military in space / Daniel Gonzales.
        p.   cm.
    "Prepared for the United States Air Force by RAND's Project AIR FORCE."
    "MR-895-AF."
    Includes bibliographical references.
    ISBN 0-8330-2661-5
    1. Astronautics, Military—United States.   2. Astronautics and civilization.   I. RAND Corporation.   II. United States. Air Force.   III. Project AIR FORCE (U.S.).   IV. Title.
UG1523. G66   1999
358 ' .8 ' 0973—dc21   98-43375
                                                                              CIP

Published 1999 by RAND
1700 Main Street, P.O. Box 2138, Santa Monica, CA 90407-2138
1333 H St., N.W., Washington, D.C. 20005-4707
RAND URL: http://www.rand.org/
To order RAND documents or to obtain additional information, contact Distribution Services: Telephone: (310) 451-7002; Fax: (310) 451-6915; Internet: order@rand.org

This report reviews the dramatic changes occurring in the commercial and foreign space sectors, the ways in which emerging commercial space capabilities can be used by the U.S. military, and what advantages adversaries may gain by exploiting those same capabilities. These changes in the space environment present the Department of Defense (DoD) with opportunities and challenges—challenges and opportunities that are examined in this report with particular reference to the space control mission area.

This report should be of interest to Air Force and DoD leaders, planners, and strategists who are concerned with the use of space assets by the U.S. military and potential adversaries, and those engaged in emerging commercial and foreign space capabilities and in the space control mission area.

This report summarizes work done in the Future Role of the Air Force in Space Project and was carried out in the Force Modernization and Employment Program of Project AIR FORCE.

PROJECT AIR FORCE

Project AIR FORCE, a division of RAND, is the Air Force federally funded research and development center (FFRDC) for studies and analyses. It provides the Air Force with independent analyses of policy alternatives affecting the development, employment, combat readiness, and support of current and future aerospace forces. Research is performed in three programs: Strategy and Doctrine, Force Modernization and Employment, and Resource Management and System Acquisition.

CONTENTS

FIGURES

TABLES

SUMMARY

The role of the U.S. military in space, particularly that of the Air Force, is changing for a number of reasons that include the growth in space capabilities outside the Department of Defense (DoD) and the severe budget pressures now being experienced by the DoD.

GROWTH IN NON-DoD SPACE CAPABILITIES

During the Cold War both Russia and the United States developed military communications satellites and other spacecraft. In the West, commercial communications satellites (Comsats) were procured by government-backed consortia, such as INTELSAT, and a few private firms. Still, the vast majority of satellites orbited during the Cold War were of military origin.

In contrast, Comsat capabilities have increased tremendously in the past decade, and civil users around the world have exhibited a growing appetite for communications. Advanced Comsats will be deployed soon in low and medium earth orbits. So many Comsats will be deployed by the turn of the century that by then most satellites on orbit will be commercial rather than military. The Comsat market is changing from a government-backed market to one dominated by private firms. Indeed, INTELSAT will soon spin off a sizable portion of its on-orbit assets to a new commercial enterprise.

During much of the Cold War, the United States and Soviet Union had a near monopoly on remote sensing satellites capable of taking high-resolution images. In the 1980s, the situation began to change, and today highly capable imaging satellites are under development by a number of countries. Russia is selling on the Internet imagery collected by previously classified satellites. Technology advances, proliferation, and market forces have blurred the dividing line between national security and commercial remote sensing systems. These trends will lead to a global commercial market for high-resolution imagery.

During the Cold War, both superpowers spent many years developing military satellite navigation systems: the U.S. Global Positioning System (GPS) and the Russian GLONASS system. As myriad civil applications for these systems became apparent, the number of civil users grew rapidly. Many domestic and foreign GPS augmentation systems now in development will provide civil users multiple ways of obtaining high-accuracy navigation services, and many will provide higher-accuracy signals than current GPS satellites. In the future, civil users, U.S. military forces, and potential adversaries will have access to high-accuracy navigation signals from a variety of sources.

EXPLOITATION OF SPACE BY ADVERSARIES

Increasing numbers of Comsats will be owned by multinational or foreign interests that may grant adversaries of the United States access to their systems. These Comsats will provide communications links to small, mobile, and, in some cases, handheld terminals. Consequently, ground terminals will be difficult to find, target, and attack using traditional means. Because a plethora of possible communication paths will be available in future Comsat systems, it may be difficult to find enemy signals in large, constantly changing satellite networks, further complicating U.S. collection strategies. For these reasons, Comsats will provide militarily significant capabilities to enemy forces, including mobile, rapidly reconfigurable, high-capacity communications links for command and control and the transmission of targeting and other intelligence information to mobile force elements.

A growing roster of countries will possess remote sensing satellites in the coming decade. More countries and transnational groups will gain access to satellite imagery produced and distributed by commercial firms, whether over the Internet or through bilateral intelligence-sharing agreements. For example, Iran could obtain satellite imagery by accessing the downlinks of Indian or Chinese satellites.

The military value of imagery depends on a number of factors. Targets imaged in archival imagery weeks or months old, may have moved or changed. The value of archival imagery depends upon target type. On the other hand, if imagery is collected and processed in near real time, it could provide near-real-time targeting and situation awareness data, including the location of mobile military targets.

The military value of imagery data also depends on satellite resolution. By the year 2000, many foreign imagery satellites will have 1- to 3-m resolution and so will be able to detect many types of militarily significant targets. Consequently, adversaries will gain strategic reconnaissance capabilities even if they have access only to archival imagery. They will be able to obtain imagery of denied areas in neighboring countries and in the continental United States (CONUS) and so will be able to target fixed facilities around the world. If adversaries obtain direct downlink access to imagery satellites and invest in processing and exploitation systems, they will obtain near-real-time imagery of the battlefield and possibly near-real-time targeting data that could be used to target GPS guided weapons.

USE OF COMMERCIAL SPACE CAPABILITIES BY U.S. FORCES

Emerging commercial space capabilities will provide U.S. forces with significant new proficiencies and can augment military systems that are in great demand, such as military satellite communications (MILSATCOM) systems.

U.S. MILSATCOM needs appear to be increasing at a rapid rate. Estimates of the aggregate MILSATCOM capacity needed to support U.S. forces in two major regional conflicts (MRCs) in 2005 range from 2.5 to 20 Gbps, a factor of up to 100 times larger than the total capacity used by U.S. forces during Desert Storm.[1] Although these estimates cannot be confirmed independently by analysis, there are at

---

[1]See Figure 4.

least three explanations for this dramatic increase in requirements: the growing use of computers on the battlefield; the need to transmit great quantities of information from CONUS to theater; and the likelihood that U.S. forces will deploy to remote regions having limited indigenous communications infrastructure.

Increasing MILSATCOM requirements present a difficult architectural issue for DoD. Early in the next decade, the entire MILSATCOM architecture will have to be replaced. Two architectural extremes bound the trade space for the next-generation architecture. At one extreme is a robust antijam (AJ) architecture designed foremost to supply jam-resistant communications. But robust AJ systems, such as MILSTAR, are expensive and deliver relatively little capacity. If only robust AJ satellites were procured, they could not satisfy projected capacity requirements.

At the other extreme is a commercial architecture that could supply large amounts of capacity using leased or purchased Comsats. In a benign environment, such an architecture may meet DoD MILSATCOM requirements. However, an adversary could easily jam such systems and degrade U.S. C2 and intelligence networks. A balanced architecture is needed that contains a mix of robust AJ and high-capacity satellites, and Comsat leases for additional surge capacity.

A factor that will affect Comsat availability is the changing structure of the international Comsat market. Private systems with foreign co-owners may dominate overseas markets and foreign owners may deny system access to U.S. forces if they object to U.S. policy. Sufficient Comsat capacity may not be available in a contingency because of capacity shortages in the spot market. This concern is especially acute in high-growth regions where Comsat capacity is in short supply. U.S. policy also limits the Comsat capacity available to U.S. forces. U.S. policy stipulates that U.S. forces can use only Comsats equipped with encrypted tracking, telemetry, and command (TT&C) links. Currently, only INTELSAT, INMARSAT, Orion, and Panamsat satellites provide coverage of overseas areas and are equipped with such links.

These issues imply that the DoD needs to develop a long-term strategy to ensure adequate and secure access to high-capacity Comsat resources. A first step was taken by the Defense Information Systems

Agency when it obtained lease options for up to 40 transponders from INTELSAT. However, additional measures will be required because the planned breakup of the INTELSAT system may significantly reduce the number of Comsats available to U.S. forces and because large amounts of additional capacity would be needed in a major contingency.

Planned U.S. commercial remote sensing satellites could provide important information services for U.S. forces. Such satellites will be able to identify militarily significant targets and could provide high-quality situation awareness, order-of-battle, and targeting information to U.S. forces, if communications links and processing resources are acquired by DoD to exploit them.

However, a large number of commercial systems are planned for an uncertain international market. U.S. commercial systems will compete against systems funded and subsidized by foreign governments and may compete against National Reconnaissance Office (NRO) systems as well, so not all planned U.S. commercial systems may survive. Thus, relying on them exclusively would not be prudent until a better understanding of the emerging marketplace is obtained. Nevertheless, it may be possible to reduce or restructure government expenditures in the remote sensing area if emerging U.S. commercial systems can establish and maintain a long-term presence in the market. The savings obtained from this reduction could be used to address emerging space control needs.

SPACE CONTROL

In the post–Cold War era, the value of space in military operations has become apparent to U.S. military planners and to potential adversaries. U.S. forces are increasingly dependent on space, especially for deployed operations overseas. Therefore, adversaries may try to deny U.S. access to space in future conflicts. Furthermore, as indicated above, emerging space capabilities can be exploited by adversaries to gain significant military advantage. Thus, space control will assume increasing importance in future military operations.[2]

---

[2]Space control is a set of activities carried out in space, on the ground, at sea, or in the air that ensure the friendly use of the space environment while denying its use to the

Defensive Space Control Measures

Defensive space control includes operational and policy measures. Operational measures consist of active and passive actions to protect space-related capabilities from adversary attack on interference. Policy measures deal primarily with treaties to outlaw ASATs. Passive actions are taken to reduce the vulnerabilities and increase the survivability of U.S. space systems and the services they provide. Specific measures may include the encryption of information transmitted to and from space systems, employment of redundant systems and service pathways, increased satellite maneuverability, and modifications to increase satellite survivability.

Several types of modifications could enhance satellite survivability. However, they increase satellite weight and complexity, and so would not be cost-effective for Comsats. Such modifications have occasionally been added to government satellites (for example, the blast shielding and high-velocity maneuver capability originally designed into the MILSTAR communications satellite). However, when the modifications were added to military satellites, procurement costs increased significantly.

During the Cold War, both the United States and the Soviet Union at different times proposed to outlaw ASATs. None of these diplomatic efforts succeeded, except to ban ASATs that employ weapons of mass destruction in space. A complete ASAT ban would be difficult to enforce and monitor, and existing Third World ballistic missiles and warheads could be modified to serve as primitive ASATs. Consequently, it is difficult to envision an effective policy solution to potential ASAT threats.

A variety of weapons and operational concepts could conceivably be used as anti-ASAT weapons or in anti-ASAT operations. For example, enemy ASAT space vehicles could be destroyed before launch by air attack. ASAT space vehicles could be destroyed in the boost phase by directed-energy weapons. Or, if an adversary stored ASATs in orbit, they could be destroyed by U.S. ASATs.

---

enemy. To accomplish these objectives, U.S. forces would have to survey space, protect the U.S. ability to use space, prevent adversaries from using space systems or services for purposes hostile to U.S. national security interests, and negate the ability of adversaries to exploit their own, foreign, or commercial space assets.

A possible near- to mid-term ASAT threat is advanced Scud-like missiles (for example, North Korean Taepo Dong missiles) used to launch many small debris objects in LEO. Satellites in orbits intersecting the debris cloud would be damaged or destroyed. A boost or midcourse ballistic missile defense (BMD) system could address this threat. A robust anti-ASAT system, i.e., one without range limitations and able to protect large satellite constellations, may have to be space-based. Such a system may resemble Brilliant Eyes and Brilliant Pebbles or a constellation of space-based lasers. This suggests that a robust anti-ASAT system could be costly to develop and controversial because of treaty and policy concerns. On the other hand, it may be possible to combine BMD and anti-ASAT functions into a single system.

Offensive Space Control Measures

Offensive space control includes operational and policy measures. Operational measures are designed to destroy or neutralize an adversary's space system or the service it provides through attacks on the space, terrestrial, or link elements of space systems. Policy measures deal primarily with agreements between the U.S. government and commercial or foreign satellite owners to deny enemy access to space services.

There are two types of policy measures intended to deny enemy access to space. The first are legally binding and apply to U.S.-owned or multinational systems, such as those contained in Presidential Decision Directive 23 (PDD23), which applies to commercial remote sensing satellites. Such measures would be effective in the most dire circumstances—for example, in a major theater war. However, they may not be effective in denying enemies access to U.S.–owned imaging satellites in lesser conflicts or crises. In particular, the PDD23 decisionmaking process could be cumbersome and prone to delays because the Secretaries of State, Commerce, and Defense must all agree to invoke the restrictions prescribed in the directive.

The second type of measure would not be legally binding or enforceable by the U.S. government. Examples of the latter include shutter control restrictions voluntarily agreed to by U.S. and foreign governments. Furthermore, it would be very difficult to verify compliance with such measures.

Destructive ASATs

Satellites in space could be destroyed by kinetic kill vehicles, or by high-power microwave or laser weapons. However, destructive ASATs have a number of drawbacks. First, they may create space debris that could damage commercial, civil, foreign, or U.S. military satellites. Second, destructive ASATs may be employable against only a very limited target set. Multinational systems may be off-limits to destructive ASATs for foreign policy reasons and because of the financial blow such attacks could inflict on the U.S. aerospace industry. Third, such ASATs could not be employed against Russian early warning or imaging satellites, nor against satellites owned by other countries in peacetime because of existing treaty prohibitions. Consequently, destructive ASATs may not be appropriate instruments of war or of deterrence in the post–Cold War era.

Nondestructive Space Control Systems

Nondestructive space control systems would prevent the functioning of target satellite payloads without damaging the target in any way. Such systems would ideally have effects that were only temporary, fully reversible, and localized or limited to a particular region on the earth's surface.

These systems are attractive for the same reasons destructive ASATs were found to be unattractive. They could be designed to not create space debris. And because they would prevent the target from functioning in only a limited region, they could be used against multinational systems. For example, nondestructive systems could enforce imagery embargoes or communications blackouts of specific areas. Such systems also may not raise as many or as serious treaty and policy concerns as destructive ASATs.

An assessment of a range of nondestructive space control system options by RAND revealed that the most-promising options worked against specific target types, i.e., Comsats or imaging satellites, but not both.

CHANGING SPACE SURVEILLANCE NEEDS

Space surveillance—the ability to detect, identify, track, and predict the position of space objects—is an essential element of space control and is necessary for providing situation awareness of the space environment.

The U.S. space surveillance network (SSN) is composed of ground-based radars, optical telescopes, communications links, and a part of the Cheyenne Mountain Operations Center (CMOC). The key product of the SSN is a real-time picture of the space environment. Because most SSN radars are located in the United States and there are a limited number of SSN telescopes, the SSN has instantaneous coverage gaps—not all space objects can be observed at any one instant in time. Consequently, accurate orbit-prediction algorithms are a necessary element in producing an accurate real-time space surveillance picture. These algorithms allow past observations to be combined with current ones to produce a space picture in which the positions of all space objects are known to a minimum position location accuracy.

Future Space Surveillance Needs

Space surveillance will likely become more important in the future because of concerns over three issues: (1) the growth in commercial space activities resulting in congestion in many orbits; (2) the growing space debris population and possible need to provide accurate warning of debris hazards to prevent the destruction of satellites; and (3) the growing importance of space in military operations and increased emphasis on future U.S. space control operations. Concerns about the changing space environment reflected in the first two issues are shared by commercial and civil space users, whereas the third issue is entirely military in origin. The preferred set of space surveillance capabilities needed to address each of these issues could be quite different and potentially could lead to different technical and organizational solutions.

Concerns have been expressed as to whether the existing SSN can adequately handle these potentially more demanding space surveillance needs—specifically, concerns have been expressed as to whether the SSN can support emerging space control needs and whether the SSN is keeping pace with technology advances.

If the coordination of international satellite operations becomes the highest-priority space surveillance task, the SSN would have to provide accurate satellite location predictions far in advance of potential collisions. A space surveillance system tailored to satisfy this need would include the most accurate orbit-prediction algorithms possible and large databases to maintain a highly accurate space catalog. This coordination activity would benefit U.S. military, civil, and commercial users, and foreign space users as well. Thus, it would seem natural to share the costs of an expanded SSN tailored to satisfy this mission. In particular, an international network of sensors could be integrated into the SSN, thereby enabling sensor costs to be shared with other developed nations that make extensive use of space.

If space debris monitoring (including debris objects with diameters as small as 1 cm), were the highest-priority space surveillance task, improved ground-based and perhaps space-based sensors would be needed. Because the small debris object population may be more than an order of magnitude larger than the current space catalog, a large state-of-the-art database would be needed. Again, a case could be made for greater burden sharing, since international civil and commercial users would benefit from a global space debris monitoring and collision warning system.

On the other hand, if the space control mission were given highest priority, space surveillance needs would depend upon the type of space control systems acquired. If nondestructive systems were developed, SSN accuracy might have to be increased by a significant factor. Not all objects in the space catalog, however, would have to be tracked at a higher level of accuracy. Only potential threats would require high-accuracy tracking. A separate high-accuracy threat track prediction system could be developed that was independent of the mainframe computers at the CMOC.

Implications for the Air Force as an Institution

The growth and emergence of new commercial space markets and the growing importance of space in military operations have led to the recognition by senior military leaders that the U.S. military must develop strategy, doctrine, and programs not only to take maximum advantage of traditional DoD and commercial space assets, but also to anticipate and prepare for the day when space itself becomes a theater of military operations. For example, the Air Force is now considering if and when to migrate certain functions, such as theater air and ground surveillance, to space. The increased use of and dependence on space by the U.S. military implied by such a migration to space will have to be accompanied by a similar internal transition within the Air Force—from an organization focused on air combat and air power to one that includes space operations as a core competency. How this organizational transition occurs and what organizational form the Air Force eventually adopts are difficult, complex questions that are beyond the scope of this investigation. However, the concrete approaches and concepts of operations that the DoD and the Air Force eventually adopt for space control, space surveillance, and battlespace surveillance missions will affect the organizational structure and the future character of the Air Force as an institution. Study of these mission areas should therefore focus not only on preferred system options and low costs, but also on organizational structures that can enable mission success and a comprehensive approach to integrated air and space operations.

ACKNOWLEDGMENTS

The author thanks his RAND colleagues Calvin Shipbaugh, Phillip Feldman, Jerry Frost, and David Trinkle, who contributed in various ways to the Future Role of the Air Force in Space project. I also thank Maj Ted Warnock, Maj Philip Sauer, and Maj Ken Verderame, who were visiting Air Force Fellows at RAND when much of this research was done, for the ideas and insights they contributed to this project.

I also thank Flora Grinage for her expert assistance in the preparation of this report.

Chapter One

GROWTH IN NON-DoD SPACE CAPABILITIES

For much of the Cold War, space was primarily the province of the two superpowers. Both sides developed imaging satellites to support military operations, gather intelligence, and verify arms control agreements. Because of their cost and space launch risks, only a relatively small number of such satellites were typically orbited at any one time, especially as these systems grew in weight, complexity, and capability.

The United States and the Soviet Union also developed military communications satellites. In the West, the first commercial communications satellites (Comsats) were owned by governments or large consortia, such as INMARSAT and INTELSAT. Governments took an active role in these consortia, assuming the risks associated with an unproven industry and supplying capital for satellite acquisition. Consortia members were either national Post Telephone and Telegraph (PTT) companies or other nationally sanctioned firms, an arrangement that enabled members to pool their resources and extend modern communications nets to less-developed regions of the world.

COMMERCIAL COMMUNICATIONS SATELLITE MARKET GROWTH TRENDS

In the post–Cold War era, many parts of the developing world are experiencing rapid economic growth and have a growing appetite for communications services of all types. Privately owned Comsats have been deployed, and in the next few years the pace of private and foreign activity in space will accelerate. The Comsat market, the most mature space market today, is undergoing a fundamental transformation from a market composed of government-sponsored consortia to one dominated by international joint ventures whose primary stakeholders are private firms. Indeed, INTELSAT will spin off a sizable portion of its on-orbit assets to a new commercial enterprise called New Skies N.V., and INMARSAT may well be pressured to do the same.

Figure 1 shows Comsat capacity estimates in terms of the number of satellite transponders available over East Asia. In 1995, Asia was on a high growth trend line, where the number of foreign-owned transponders was tripling about every five years. If this pattern continues, there will be over 1800 foreign-owned transponders over Asia by 2000.

Other parts of the developing world are on a nominal growth path, where the number of transponders is doubling every five years. In the more mature markets of Europe and the United States, where a large number of satellites are already on orbit, there is a shortage of orbital slots and significant competition from entrenched ground-

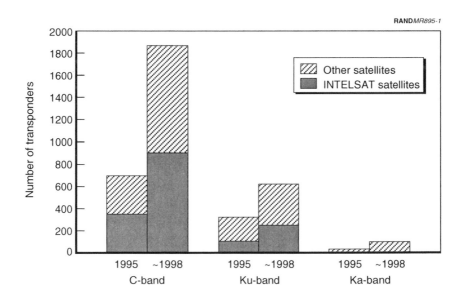

Figure 1—Comsat Transponder Growth Trends for East Asia

based networks.[1]   These mature markets are on a low growth path, where the number of transponders is growing by perhaps only 25 percent every five years.

In the next five years, the Comsat market will rapidly expand and provide a range of new communications services. Constellations of advanced Comsats, many with onboard circuit switching or networking capabilities, will be deployed for the first time in low earth orbit (LEO) and medium earth orbit (MEO).

Low Data Rate Systems

The simplest of the new systems are called "Little LEOs." The general characteristics of two leading Little LEO systems are shown in Table 1. The systems will employ small satellites and provide low data rate services such as paging and electronic mail to users equipped with appropriate handheld devices. Users will transmit short messages to the satellites when they are overhead. The satellite will store messages and forward them when the satellite comes in view of the recipient. Little LEO users will experience communication time delays of minutes to hours because of the time needed for satellites

Table 1

Planned LEO Low Data Rate Systems

System	Orbit	Frequency Band	Principal Contractor	Number of Satellites	IOC/FOC
ORBCOMM	LEO	VHF/UHF	Orbital Sciences Corp.	36	1995/1999
LEO One	LEO	VHF/UHF	LEO One Corp.	48	?

SOURCES: Robert Ropelewski, "Satellite Services Soar," *Aerospace America,* November 1996; Technical Description of ORBCOMM Communication System and Spacecraft, ORBCOMM, December 1994.

NOTES: VHF = very high frequency, UHF = ultra high frequency, IOC = initial operational capability, FOC = full operational capability.

---

[1]Traditional Comsats operate in geosynchronous orbit (GEO) and provide bulk circuit transmission services over high data rate (HDR) links. To prevent signal interference, satellites in GEO are positioned a minimum distance apart. Consequently, these positions, or orbital slots, are a scarce resource.

to move within view of recipients. However, Little LEO services and end-user equipment are expected to be quite inexpensive relative to traditional Comsat services.

The pioneer in this market is ORBCOMM. A full constellation of 36 ORBCOMM satellites will be launched in 1999. Other systems have been proposed, but their development is questionable, either because of licensing or financing concerns.[2]

LEO and MEO Voice Systems

Constellations of larger LEO and MEO satellites, also planned, will provide voice and medium data rate (MDR) communications of up to 64 kbps to users on a worldwide or near-worldwide basis.[3] The two pioneering systems are Iridium and Globalstar. These constellations will contain a large number of satellites (see Table 2). Iridium satellites will have sophisticated onboard processing capabilities and satellite crosslinks, whereas Globalstar will use ground stations to route and switch voice calls. All the voice systems in Table 1 will provide links to handheld cellular telephone-like terminals.

Table 2

Planned LEO and MEO Voice Systems

System	Orbit	Frequency Band	Principal Contractor	Number of Satellites	IOC/FOC
Indium	LEO	L&S	Motorola	66	1998/1999
Globalstar	LEO	L&S	Loral	48	1998/1999
Ellipso	LEO, MEO	L&S	Boeing	17	2001/
Ecco	LEO	L&S	Orbital Sciences	12+35	2001/
ICO Global	LEO	L&S	Hughes	12	1999/

SOURCES: Neal Hulkower, *Update on the "Big LEOs,"* MITRE, Bedford, Massachusetts, March 7, 1995; *Iridium Today,* Iridium Inc., Washington D.C., Fall 1994; Robert Ropelewski, "Satellite Services Soar," *Aerospace America,* November 1996. "Boeing Wins Ellipso in Satcom Shuffle," *Aviation Week and Space Technology,* May 11, 1998; p. 35.

---

[2]Another Little LEO, General Electric/Americom, has received a Federal Communications Commission (FCC) license to operate but has reportedly had trouble raising financing.

[3]Only Iridium will provide full-earth coverage. The other voice systems will not provide communications in the polar regions.

High Data Rate Bandwidth-on-Demand (BOD) Systems

The systems shown in Table 3 will provide HDR communications, or multimegabit-per-second links, to users equipped with very small mobile terminals with 1- to 3-ft antennas.[4] They will have sophisticated onboard processing capabilities and provide HDR links for video teleconferencing, high-speed Internet access, and other forms of data communications. These services will eventually be available over all landmasses and some ocean areas. They will also have the important advantage of short link-setup times, i.e., they will provide BOD. Today it can take days, weeks, or even months to set up a Comsat link, especially in foreign countries. In contrast, BOD systems will provide circuit-setup times on the order of minutes and perhaps even seconds to users with previously established system accounts.

Table 3

High Data Rate Bandwidth-on-Demand Satellite Systems

System	Orbit	Frequency Band	Principal Contractor	Number of Satellites	IOC/FOC
Spaceway	GEO	Ka	Hughes	17	1998/2000
Cyberstar	GEO	Ka	Loral	3	1999/2003
Astrolink	GEO	Ka	Lockheed-Martin	9	1999/2003
PANAMSAT	GEO	Ka	?	10	1999/?
SkyBridge	LEO	Ka	Alcatel	80	?
Teledesic	LEO	Ka	Motorola	288	2002/?

SOURCES: Robert Ropelewski, "Satellite Services Soar," *Aerospace America,* November 1996; "Europe Testing Technologies to Keep Pace With U.S.," *Aviation Week and Space Technology*, March 31, 1997, p. 51.

REMOTE SENSING SATELLITE MARKET GROWTH TRENDS

During much of the Cold War, the United States and Soviet Union had a near monopoly on remote sensing satellites, and in particular

---

[4]As opposed to the larger terminals needed for two-way HDR communications links with traditional systems.

on systems capable of taking high-resolution images or photographs of the earth's surface. In the 1980s, the situation began to change.

U.S. Civil and Commercial Remote Sensing Satellites

During the 1980s, the United States and France separately developed the first civil remote sensing satellites, LANDSAT and SPOT. These were relatively large and expensive satellites. Efforts to privatize LANDSAT failed because it could produce only black and white (panchromatic) images with 30-m resolution. In contrast, SPOT had a panchromatic resolution of 10 m. By the late 1980s, SPOT had captured the largest share of the commercial satellite imagery market, and during the Gulf War the U.S. military made extensive use of SPOT imagery. Shortly after the Gulf War, Russia began to market high-resolution satellite photos from previously classified satellites, and other countries such as India and Israel accelerated development of highly capable imaging satellites. It also soon became apparent that an expanding number of countries would be entering the satellite imaging market with systems significantly more capable than LANDSAT.

Permission was given to U.S. firms to develop, sell, and operate remote sensing satellites with resolutions of up to one meter in Presidential Decision Directive (PDD) 23, which was signed by President Clinton in March 1994. PDD 23 has been controversial in some quarters because it is seen to encourage the proliferation of high-resolution imaging satellites and imagery, products that could threaten the security of the United States and its allies, such as Israel.[5] On the other hand, PDD 23 is also viewed, especially in the U.S. aerospace industry, as an attempt to "level the competitive playing field" and enable U.S. firms to compete unencumbered against foreign firms in a potentially important emerging market that could eventually grow to over $20B a year in imagery sales.[6] If U.S. firms were prohibited from entering this market at the same time that defense spending and DoD procurement budgets are falling

---

[5]"Israel Wants Imagery Ban," *Space News*, June 17–23, 1996, p. 1.

[6]*Emerging Markets of the Information Age: A Case Study in Remote Sensing Data and Technology*, C. B. Gabbard, K. O'Connell, G. S. Park, and P.J.E. Stan, Center for Information Revolution Analysis, RAND, DB-176-CIRA, January 1996.

precipitously, U.S. firms could not only lose their technological advantage in this area, but might also have to abandon the market altogether, a prospect that eventually could have serious national security implications.

Table 4 shows the characteristics of the commercial 1-m-resolution imaging satellites proposed by U.S. contractors in response to PDD 23. Ikonos, Earth Watch, and ORBIMAGE, with the help of foreign equity partners, are developing such satellites. At least two of these systems will have multispectral and stereoscopic imaging capabilities, and all will be able to determine the location of imaged objects with a fair degree of accuracy. Most of these systems will be in sun-synchronous orbit and will have a 2- to 5-day revisit. The potential exception is the Earth Watch satellite, Quick Bird, which may be placed in a 50-deg inclined orbit, where it could revisit locations at mid latitudes twice a day.

A few other remote sensing systems may also be built by U.S. manufacturers. A Boeing-led venture may build a system called Resource 21. Motorola has reportedly been granted a remote sensing license,

Table 4

Characteristics of U.S. Commercial High-Resolution Imaging Satellites

Satellite	Resolution (m)[a] p	ms	Scan Line Width (km)	Geolocation Accuracy (m)	Orbit Attitude (km)	Orbit Inclination (deg)
Ikonos						
S1S1, S1S2	1	4	11–15	10–14	680	98.1
Earth Watch:						
Quick Bird	1	4	15–37	2–5[b]	470	48–52
ORBIMAGE:						
ORBVIEW #3	1	4	15	10–14	400	98.2
ORBVIEW #4	1	8	15	10–14	700	98.2

SOURCES: "New Satellite Images for Sale," *International Security*, Vol. 20, No. 1, Summer 1995, pp. 94, 125; http://www.spaceimage.com; and *Proceedings of the Land Satellite Information in the Next Decade Conference*, American Society of Photogrammetry and Remote Sensing, Vienna, Virginia, September 1995.

[a]p = panchromatic; ms = multispectral.

[b]Requires upgrade of satellite navigation package.

although it has yet to announce specific plans. A number of other remote sensing satellites with advanced multispectral imaging sensors are being developed by U.S. firms under NASA sponsorship. These could lead to additional commercial ventures in the coming decade.

Foreign Remote Sensing Satellites

Table 5 gives a chronology of foreign remote sensing satellite developments. The progressive improvement in the capabilities of the French SPOT system are shown in the first row. SPOT 3 and 4 have 10-m resolution. The next-generation satellite, SPOT 5, will have 2.5-m resolution.

The European Space Agency (ESA) has orbited a series of synthetic aperture radar (SAR) satellites for environmental research. ESA has plans for future remote sensing satellites, but detailed specifications for these satellites and launch dates were unavailable when this research was conducted.

Canada has developed a highly capable civil SAR satellite, RADARSAT, which provides 10-m resolution. There have been discussions about privatizing RADARSAT, and the Canadian contractor, Spar Aerospace, has had discussions with Chinese officials concerning the sale of one or two RADARSAT-like satellites to China. NASA officials have also explored the possibility of supporting development of SAR satellites with commercial and civil applications.

Russia has commercialized a number of what were previously military surveillance satellites and offered imagery from these systems for sale on the commercial market. SAR imagery is available from ALMAZ with 10- to 15-m resolution. The next ALMAZ satellite, ALMAZ-1B, if it is developed, will carry a number of sensors, including a 5-m resolution SAR and 2.5m panchromatic visible light camera system capable of electronically downlinking imagery to the ground.

Today one can obtain Russian KVR-1000 satellite imagery with 1-m resolution from a number of distributors, including Microsoft. The KVR-1000 satellite is a film-based system where film is returned by capsule, which can lead to significant image processing delays.

Table 5

Current and Planned Foreign Remote Sensing Satellites

Country	Satellite				
	1985	1990	1995	2000	2005
France	SPOT 1 10 m	SPOT 2 10 m	SPOT 3 10 m	SPOT 4 10 m	SPOT 5 2.5 m
ESA		ERS-1 SAR 30 m		ERS-2 SAR 30 m	ENVISAT SAR, EO
Russia		ALMAZ SAR 10–15 m	KVR-1000 2 m	ALMAZ-1B SAR 5 m, 2.5 m	KVR-1000 1 m
Canada				RADARSAT SAR 10 m	RADARSAT-2 3 m
China	CHINASAT 1/2 80 m			CBERS1&2 CBERS 3&4 20 m 5 m	Unnamed system 2.5 m
Brazil			MECB SSR1 200 m		
India	IRS-1A 36 m	IRS-1B 36 m	IRS-1D 5 m	CARTOSAT-1 2.5 m	CARTOSAT-2 0.5–1.0 m
Japan (Civil)	MOS-1 50 m	MOS-1B 50 m	JERS-1 SAR 25 m 18 m	ADEOS 8 m	ALOS 2.5 m
Military systems[a]			Helios ~1 m EO	Osiris (France) SAR Japan 1 m EO, SAR	Aus (P 2044) <1 m EO

SOURCES: *Proceedings of the Land Satellite Information in the Next Decade Conference*, American Society of Photogrammetry and Remote Sensing, Vienna, Virginia, September 1995; *Proliferation of Satellite Imaging Capabilities: Developments and Implications*, Berner Lanphier and Associates, Inc., February 27, 1995; Robert Ropelewski, "Satellite Services Soar," *Aerospace America*, November 1996.

NOTE: Other emerging capabilities are from South Korea, Israel, Pakistan, South Africa, Taiwan, Argentina, United Arab Emirates, Thailand. EO: electro-optical (unless otherwise specified, EO sensors are implied).

[a]Chinese and Russian military systems not shown.

As indicated in Table 5, China has an active remote sensing satellite program. China and Brazil are jointly building a pair of remote sensing satellites, the China Brazil Earth Resources (CBERS) satellites. CBERS 1 and 2 will have 20-m resolution, and the first satellite was scheduled for launch in 1997 or 1998. China has expressed interest in developing with Brazil additional CBERS satellites with improved resolution. China has also announced the intention to develop one or more "civil" remote sensing satellites with 2.5 m or better resolution.

India has an active remote sensing satellite program. It has a system in orbit with 6-m resolution, and plan to launch systems with 1- and 2.5-m resolution in the next five years. Indian satellite imagery is sold on a commercial basis around the world.

Japan also is progressively developing more-capable remote sensing satellites. Its current civil system, the Advanced Earth Observation System (ADEOS), has 8-m resolution, and its next-generation system, the Advanced Land Observation Satellite (ALOS), will have a resolution of 2.5 m. According to recently published reports, the Japanese military will orbit two visible imaging and two radar imaging satellites. These four satellites have 1-m resolution.[7]

Finally, as indicated in the table, a number of other countries besides those mentioned have or may develop remote sensing satellites specifically for national security purposes. France, Spain, and Italy launched the first European military surveillance satellite in 1996, Helios-1A. The French and Germans have had extended discussions concerning development of the next-generation Helios satellite and a military SAR system. There have been reports in the Australian press concerning development of national high-resolution imaging satellite capabilities. Israel recently launched OFEC-3 which may have been developed as a military surveillance system but now may be privatized.

In summary, highly capable imaging satellites are under development or are being used for commercial purposes by a number of countries. Technology advances and proliferation have blurred the

---

[7]"Japan plans $1.3 billion Spysat program to counter N. Korea," *Aerospace Daily*, Vol. 188, No. 25, November 4, 1998, p. 193.

dividing line between national security and commercial or civil re-mote sensing systems. U.S. commercial systems will have to com-pete against systems funded and subsidized by foreign governments and may compete with the National Reconnaissance Office (NRO) for U.S. contracts if the policy framework separating the commercial and government remote sensing communities is not clarified. Re-gardless of how these issues are resolved, these trends will lead to in-creased availability of high-resolution imagery and to a worldwide commercial market for such imagery in the near future. How big this market may be and how many players it may support are the subjects of considerable debate.[8]

SATELLITE NAVIGATION SERVICES

The Global Positioning System (GPS) is a constellation of 24 MEO satellites. Each satellite transmits timing and location data to users equipped with small receivers. GPS receivers use information from multiple GPS satellites to determine the receiver's position to a high degree of accuracy. The current generation of GPS satellites provides about 60-m horizontal positioning accuracy, or circular error prob-able (CEP), to commercial users using the coarse acquisition (CA) code and 10-m CEP to military users who have access to the preci-sion (P) code. The next generation of GPS satellites, the Block 2R system, will deliver 20-m accuracy to commercial users using the CA code and 6-m CEP to military users having access to the P code.

A number of U.S. civil government agencies are planning to augment GPS to provide navigation services. A GPS augmentation system takes GPS signals, determines corrections to them, and broadcasts these corrections as a secondary signal (i.e., a differentially corrected signal). Such differential GPS (DGPS) systems can deliver high-accuracy position location information even though they use the unencrypted CA code. Wide-area DGPS systems will deliver position location accuracies from 15 to 1.5 m.

The U.S. Federal Aviation Administration (FAA) is developing wide-area and local-area DGPS systems. The local-area systems will pro-vide guidance information for precision landing at airports at accu-

---

[8]Gabbard and O'Connell, et al., 1996.

racy levels of on the order of one to one third of a meter. The FAA Wide Area Augmentation System (WAAS) will provide 15-m accuracy.[9] The Coast Guard is developing a DGPS system for maritime users that will provide location accuracy of 1.5 m. Other government agencies also are planning and developing similar systems.

Other GPS augmentation systems that could provide high-accuracy navigation services are under development or study overseas. DGPS systems are already in operation in some European countries. Prototype systems are being developed in China and Poland, and are planned for South Africa, the Middle East, and other regions. Wide-area DGPS systems are also planned in Europe and in Asia to provide high-accuracy navigation signals to civilian airliners and other users via Comsats.

Many domestic and foreign GPS-related systems in development will provide commercial and civil users multiple ways of obtaining high-accuracy navigation services. In addition to the proliferation of DGPS systems, the Russian GLONASS constellation of satellites provides navigation services with accuracy comparable to GPS. The Russians have publicly stated that they will maintain GLONASS in spite of the financial difficulties faced by their space industry. European countries such as Germany have explored joint ventures in space navigation with Russia to reduce their own dependence on the U.S.-controlled GPS system. Therefore, it is probable that commercial users, U.S. military forces, and potential adversaries will have access to high-accuracy navigation signals from a variety of sources.

---

[9]S. Pace, G. Frost, I. Lachow, D. Frelinger, D. Fossum, D. K. Wassem, and M. Pinto, *The Global Positioning System—Assessing National Policies*, Critical Technologies Institute, RAND, MR-614-OSTP, 1995.

THE IMPACT OF EMERGING SPACE CAPABILITIES ON MILITARY OPERATIONS

In Chapter Two, we examine the military implications of emerging space systems for U.S. forces and for potential adversaries.

EXPLOITATION OF SPACE BY ADVERSARIES

A growing roster of countries will possess their own remote sensing satellites in the coming decade. Even more countries and groups will have access to advanced Comsat services and satellite imagery.

Communications Satellites

Increasing numbers of Comsats will be owned by foreign interests who may be willing to grant U.S. adversaries access to their systems. Political or diplomatic pressures from the U.S. government may be ineffective because of profit objectives or differing views on U.S. policy. Thus, adversaries may be able to use Comsats for command and control (C2) and intelligence dissemination purposes.

The ever increasing technological capabilities of Comsats are another of concern. Today, many foreign entities are procuring very small aperture terminals (VSATs) that are highly mobile. VSATs can be mounted on vans or trucks, making them ideal for the C2 of mobile force elements such as Scud missile launch units. In a few years, anyone will be able to purchase hand-held Iridium, Globalstar, or ICO voice terminals that can be concealed by agents or special forces and mounted on aircraft, ground vehicles, ships, and submarines.

The trend toward VSATs and hand-held terminals has significant implications for the U.S. military. A major component of information warfare strategy is counter-C2, with the purpose of destroying or degrading enemy C2 networks. Traditional ways to destroy enemy C2 networks rely upon the ability to detect, locate, and attack key communications nodes in such networks. In the past, such nodes could be attacked with strike aircraft because they were usually fixed or transportable assets. However, if adversaries decide to use advanced Comsats with VSATs or hand-held terminals, it would be difficult to target and attack such C2 nets using conventional means. Indeed, one lesson of the Gulf War is that unit mobility can significantly enhance unit survivability against air attack.

It may also become more difficult to jam enemy communications on advanced Comsats, especially BOD systems such as Iridium, Spaceway, or Teledesic. An adversary could access the system at a specific frequency and over a specific satellite for a short period of time and then terminate access. A short time later the adversary could regain system access at a different frequency using a different satellite in the same system constellation.

Because of the plethora of possible communication paths available in future Comsat systems, it could be difficult to find enemy signals in these large and constantly changing satellite networks. And if U.S. forces used the same system and tried to jam an adversary's signals, it could cause significant collateral damage to U.S. nets or to commercial or neutral third-party users of the same system.

In summary, Comsats can provide militarily significant capabilities to enemy forces. Future Comsats could provide mobile, rapidly reconfigurable, high-capacity communications links for C2, and the transmission of targeting and other intelligence information, especially to mobile force elements. Adversaries exploiting emerging HDR BOD systems could acquire robust and flexible intelligence dissemination networks that could be reconfigured in near real time, making it difficult for U.S. strike forces to target and destroy these networks by conventional means.

Remote Sensing Satellites

Adversaries will be able to access and use remote sensing satellites. Chapter One indicated that potential adversaries will gain access to high-resolution imagery in the coming decade, sometimes with significant time delays if the imagery is obtained from archives, but in other circumstances in near real time if direct downlink access to imaging satellites is available. Access to such imagery could be gained by commercial means or from intelligence-sharing agreements with countries possessing imaging satellites. For example, a country like Iran could obtain satellite imagery by accessing the satellite downlinks of systems owned by India or China.

Direct satellite access may be included in future commercial contracts. Imaging-satellite downlink stations are being equipped so they can acquire and interpret downlinks from multiple satellites, as imagery distributors try to broaden their product lines. Thus, a single terminal may be used to access a number of imaging satellites. An example of the proliferation of downlink stations is illustrated in Figure 2, which shows the 20 SPOT downlink stations currently in operation around the world.

RAND*MR895-2*

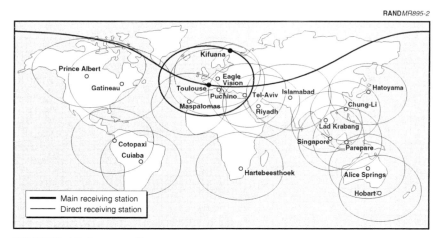

SOURCE: http//www.spotimage.com.

Figure 2—SPOT Downlink Receiving Stations

What is the intelligence value of information acquired by these systems? Figure 3 shows the sensor resolution needed to detect and generally or precisely identify a number of targets such as military ground units, vehicles, and aircraft. For example, with a resolution of 6 m one can detect the presence of aircraft at an airbase. With a resolution of 1.5 m one can generally identify the type of aircraft or distinguish large from small aircraft, i.e., a fighter from a transport plane. With 0.9-m resolution, one can precisely identify aircraft or distinguish an F-15 from an F-16. Similar object identification estimates are shown in the figure for other target classes. We note that the resolution requirements shown in the figure are approximate and depend on specific observation conditions, such as sun angle, observation angle, height of the target, and target color.

The data in Figure 3 apply to ordinary "raw" imagery, i.e., with no image enhancement, no hyperspectral subpixel image enhancement, and no special stereoscopic image processing. Image enhancement techniques can reduce image resolution requirements for target detection and identification by a factor of two or more in some cases.[1] However, even without advanced image processing techniques, by the year 2000, when many foreign systems will have 1- to 3-m resolution, adversaries will likely have the capability to detect and distinguish many types of militarily significant targets.

If an adversary possessed its own imaging satellites or had direct downlink access to those of a foreign power, and had made the necessary investment in processing, exploitation, and dissemination systems (and in people), it could obtain near-real-time imagery of the battlefield. And if the adversary can exploit next-generation Comsats to deliver imagery data to its military forces on the battlefield, its own battlefield awareness and targeting capabilities would be greatly enhanced.

However, adversaries could be hampered by the same problems that plagued U.S. forces during the Gulf War. They could experience significant processing and transmission delays in delivering imagery

---

[1] For example, particular types of trees can be distinguished using multispectral or hyperspectral sensor data, even though such trees cannot be distinguished in an ordinary visual image of the same scene.

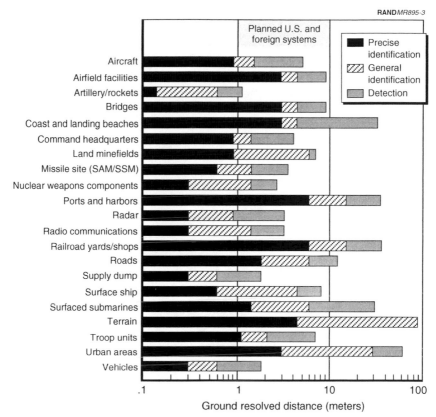

SOURCE: *Tactical Satellites*, briefing to the NORAD Wide Area Surveillance Conference, Rockwell Aerospace, April 1995.

Figure 3—Target Detection and Identification from Space

data to military commanders, thus limiting the tactical utility of the data. Whereas mobile targets could be detected, they might move after the image was taken and targeting data supplied to missile launch units. Thus, mobile targets might not be vulnerable to attack in this case.

Nevertheless, even if an adversary encounters significant time delays in processing and disseminating imagery data, it could still use the data to target fixed facilities. For example, an adversary could identify and target key weapons production facilities in adjacent countries.

In summary, in the next decade many countries will have militarily significant satellite imaging capabilities. An even larger number of countries will have access to militarily significant satellite imagery from either the emerging commercial satellite imaging market or intelligence-sharing agreements with foreign powers. Consequently, many countries will gain strategic reconnaissance capabilities. They will be able to image denied areas in neighboring countries and in the United States, and target fixed facilities around the world.

Just as with Comsats, some traditional U.S. countermeasures may not be effective in preventing access to future imaging satellites. For example, the downlink terminals used to receive imagery may be difficult to locate if they are mobile, and they could be difficult to jam from the ground if they were located deep inside enemy territory.

USE OF COMMERCIAL SPACE SYSTEMS BY THE U.S. MILITARY

Emerging commercial space systems could provide U.S. forces with significant new capabilities and could augment military systems that are in great demand, such as military satellite communications (MILSATCOM) systems.

Communications Satellites

The U.S. military's need for satellite communications is increasing at a rapid rate. Shown in Figure 4 are projected DoD aggregate MILSATCOM requirements for two major regional conflicts (MRCs) for about the year 2005. Also shown as a benchmark is twice the capacity used by U.S. forces during Desert Storm.[2] Estimates of the capacity needed for two MRCs from different requirements studies range from 2.5 to 20 Gbps, a sum that is up to a factor of 100 larger than that used by U.S. forces during Desert Storm.

---

[2]RAND research on Desert Storm indicated U.S. forces used about 100 Mbps of capacity. About 75 percent was supplied by military satellites. The rest was supplied by Comsats because additional MILSATCOM capacity was not available.

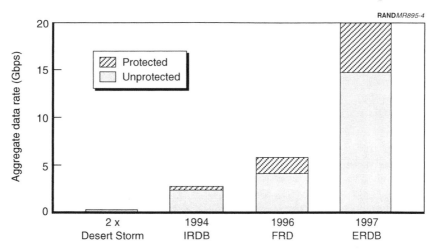

SOURCES: Leland Joe and Daniel Gonzales, *Command, Control, Communications, and Intelligence Support of Air Operations in Desert Storm*, RAND, N-3610/4-AF, 1993; *Communications Mix Study Progress Report*, MITRE, Reston, Virginia, February, 1997.

NOTE: IRDB = Integrated Requirements Database, FRD = Function Requirements Database, ERDB = Emerging Requirements Database.

Figure 4—MILSATCOM Requirements Growth for Two Major Regional Conflicts

There are a number of reasons why these requirements appear to be increasing so dramatically—the growing use of computers on the battlefield; increasing emphasis on transmitting large quantities of information from CONUS to military forces in the theater; and the likelihood that U.S. forces will have to fight in remote regions of the world where a modern and highly capable communications infrastructure may not be available.

It is difficult to determine how much communications capacity U.S. forces will need in future conflicts. During Desert Storm, there were significant capacity shortages. A great deal of intelligence information was not electronically transmitted to the theater because of insufficient bandwidth.[3]   Difficulties were also encountered in transmitting air tasking orders (ATO) and other information to military units. Because of the increasing demands for data communications at all echelons, it is unlikely that MILSATCOM systems alone will be able to supply all the capacity needed in future conflicts.

Increasing MILSATCOM requirements present a significant architectural issue for DoD and the Air Force. Early in the next decade, the entire MILSATCOM architecture will have to be replaced. Two architectural extremes bound the trade space for the next-generation MILSATCOM architecture. At one extreme is a robust antijam (AJ) architecture, or one designed foremost to supply survivable and fully jam-resistant communications. The most jam-resistant system today is MILSTAR, but these satellites cost over a billion dollars each, and deliver relatively little capacity compared to present day Comsats. Consequently, only a small number of robust AJ satellites can be afforded as replacements for existing systems. If only robust AJ satellites were procured, they could not supply the capacity needed to support wartime surge capacity requirements to deployed forces.[4]

At the other extreme of the trade space is a commercial off-the-shelf (COTS) system architecture that could supply large amounts of communications capacity using satellites designed to operate at military frequencies. In a benign environment, this architecture could meet DoD MILSATCOM capacity requirements. However, an adversary could easily jam such systems and potentially degrade U.S. C2 and intelligence networks.

The optimal architecture is one in which a balance is struck and a suitable mix of robust AJ and high-capacity satellites is procured, and in which Comsats are leased to provide additional surge capacity. An integrated DoD investment and leasing strategy is needed to secure large amounts of Comsat capacity and to ensure its availability during wartime.

---

[3]Joe and Gonzales, 1993.

[4]Laser communications systems could potentially supply high-capacity AJ links. However, it is unclear whether such systems would be able to provide an all-weather high-capacity capability to U.S. ground forces or Navy ships and submarines.

Comsat Availability. A factor that will affect Comsat availability to U.S. forces is the structure of the international Comsat market. The market was once dominated by INTELSAT and INMARSAT, but their control over the market is fading rapidly. Private systems with foreign owners will likely dominate the international Comsat marketplace, and in particular foreign GEO Comsat markets. Foreign Comsat owners or shareholders may deny system access to U.S. forces if they object to U.S. military or foreign policy objectives or if they wish to remain neutral in a conflict. Although foreign shareholders cannot deny U.S. forces system access in the United States or allied countries, international frequency allocation agreements may allow them to deny system access in their own country, and they may influence neighboring countries to do the same. Thus, U.S. access to Comsats that provide coverage in the conflict area may not be ensured in some cases.

A second reason why sufficient Comsat capacity may not be available to U.S. forces in a contingency derives from the changing nature of the market. As the market transitions from one dominated by INTELSAT and INMARSAT to one composed of private competitive firms, market efficiency may increase, and the percentage of total on-orbit Comsat capacity available in the spot market may decrease significantly. Private owners are less inclined to leave transponders empty, and are more willing to lower prices to stimulate market demand. Thus, the Comsat spot market may not be able to provide large amounts of capacity to U.S. forces on short notice. This concern is especially acute in high-growth regions of the world where Comsat spot-market capacity is already in short supply.

A third reason why sufficient Comsat capacity may not be available to U.S. forces in a contingency results from current U.S. policy regarding the use of foreign-owned Comsats. Current U.S. policy stipulates that U.S. forces cannot use Comsats that are not equipped with a U.S.-government-approved encryption capability for the satellite's tracking, telemetry and command (TT&C) links. The only Comsats today that provide overseas coverage and are equipped with such links are INTELSAT, INMARSAT, Orion, and PANAMSAT. The

Comsat options available to U.S. forces in future conflicts could be reduced significantly unless there is a change in policy.[5]

These issues imply that the DoD needs to develop long-term measures to ensure access to Comsat capacity sufficient for major contingency operations. A first step was taken by the Defense Information Systems Agency (DISA) when it obtained an option for up to 40 transponder leases from INTELSAT. Additional measures will be required, however, especially because the planned breakup of the INTELSAT system may eventually reduce the number of Comsats available to U.S. forces, either because of TT&C policy restriction or because unfriendly foreign interests may become owners of on-orbit assets formerly owned by INTELSAT.

One other option is to transition fixed DoD backbone networks, such as the Air Force satellite control network or the Defense Information System Network (DISN), to Comsats and, if possible, to procure backup fiber optic leases for these backbone networks. By transitioning backbone networks to Comsats and having the option to move them to fiber-optic links in a contingency, the offloaded Comsats could be used to provide surge capacity to U.S. forces. This suggests that some combination of Comsat leasing and fiber "CRAFing," if appropriately contracted for, could make a substantial amount of Comsat capacity available to U.S. forces in a contingency.[6]

A second option is for the DoD to procure Comsats as COTS products, and to launch and store them on orbit or to store them on the ground and launch them when needed. However, as mentioned above, relying on COTS Comsats may be putting U.S. forces at risk because of jamming threats.

---

[5]At the time this research was completed, it was not known whether planned systems such as Spaceways, Cyberstar, or Teledesic would be equipped with encrypted TT&C links. Even though Iridium is not equipped with encrypted TT&C links, the Army has purchased an Iridium gateway and a large number of Iridium handsets. By maintaining control of the gateway and the call setup and tear-down process, the Army will be able to establish secure voice links on Iridium.

[6]The Civil Reserve Air Fleet (CRAF) is a mechanism by which the DoD can lease U.S. commercial aircraft in times of national emergency. Fiber "CRAFing" would make fiber-optic networks available to the DoD at preset lease prices.

A third option is for the DoD to develop surrogate satellites, such as unmanned aerial vehicles (UAVs), equipped with communications payloads that could provide sufficient capacity to U.S. forces on the battlefield. However, the use of UAVs as communications relays presents a host of new vulnerability, network control, and network assignment issues that have yet to be resolved.

Further analysis is needed to determine the best DoD strategy for ensuring access to sufficient satellite communications capacity in wartime.

Remote Sensing Satellites

As indicated in Figure 3, 1-m U.S. commercial remote sensing satellites will be able to identify militarily significant targets. These commercial systems can provide high-quality situation awareness, and order-of-battle and targeting information. However, because a large number of commercial systems are planned for an uncertain international market, U.S. commercial systems will have to compete against systems funded and subsidized by foreign governments. Consequently, not all planned U.S. systems may survive over the long term. Thus, relying on them exclusively would not be prudent until a better understanding of the emerging imagery marketplace is obtained.

In addition, the tactical utility of these systems will depend upon a number of important details—how quickly they can be tasked, how responsively imagery can be delivered, and how flexibly they can be retasked against new targets. Answers to these questions depend on the technical features of the systems and how high a priority U.S. military commercial imagery orders are given relative to those of other commercial clients.

It is expected that a huge quantity of imagery will be produced by these systems. DoD will have to expand its own imagery exploitation and dissemination capabilities to deal with a flood of new information, or it will have to outsource these functions while ensuring that imagery products can still be quickly delivered to deployed forces in the theater.

Emerging U.S. high-resolution commercial remote sensing systems are narrow-field-of-view or "soda straw" systems. They will frequently need to be cued either by other systems, such as LANDSAT or SPOT, to image point targets of interest. Therefore, from a military standpoint, they should not be considered as replacements for LANDSAT and SPOT, which provide substantial broad-area coverage capability, even though in the commercial markets they are likely to compete directly against LANDSAT and SPOT. An important issue is whether the United States should continue to develop and operate a LANDSAT type of system.[7] We believe there are valid national security reasons to maintain a U.S. broad-area surveillance capability.

_____

[7]LANDSAT 7 is currently funded as part of the NASA Earth Observation System (EOS) program. However, its funding status has frequently been in question throughout its history.

Chapter Three

THE SPACE CONTROL DEBATE

The debate over the military utility of space systems and potential U.S. responses to space systems' use by an adversary has waxed and waned for some time. Below we review some of the historical arguments both for and against space control and examine how the present debate differs from the one that raged during the Cold War.

SPACE CONTROL

Space control is a set of activities that could potentially be carried out by U.S. forces in space, on the ground, at sea, or in the air, to ensure the friendly use of the space environment while denying its use to the enemy. To accomplish these objectives, U.S. forces would have to survey space, protect the U.S. ability to use space, prevent adversaries from using space systems or services for purposes hostile to U.S. national security interests, and negate the ability of adversaries to exploit their own, foreign, or commercial space assets.

Counterspace operations would be missions carried out to achieve space control objectives either by defensive or offensive means. A destructive antisatellite (ASAT) interceptor weapon that destroys the target satellite on impact is an example of an offensive capability that could be developed by the United States. Adding high-thrust rockets to an existing military satellite in order to give the satellite the ability to maneuver away from ASAT threats is an example of a defensive space control capability.[1]

THE COLD WAR SPACE CONTROL DEBATE

The space systems of both superpowers played an important but often hidden role during the Cold War. Systems such as the Defense Support Program (DSP) were sources of early warning data. The original purpose of DSP was to provide high-confidence warning of a Soviet nuclear attack to the National Command Authority (NCA) as early as possible so that an appropriate response could be formulated.

Communications satellites provided worldwide connectivity from the NCA to forward-deployed strategic force elements, enabling assured and deliberate control of these elements by the highest levels of government.

Reconnaissance satellites monitored arms production and storage facilities and the status of military forces deep inside the Soviet Union. These capabilities, also referred to as national technical means (NTM), were used to assess the composition and magnitude of the Soviet threat and to verify that the Soviets complied with arms control agreements. Indeed, the role of NTM was explicitly referred to in arms control agreements, indicating the importance of NTM in arriving at and in maintaining such agreements.

Because space systems provided valuable information and connectivity, all three types of systems were viewed as stabilizing influences on the strategic balance during the Cold War. Accurate information on enemy intentions and capabilities and assured connectivity to dispersed strategic force elements were viewed as vital to avoid miscalculations on either side and to prevent the accidental launch of nuclear weapons.

---

[1]*U.S. Space Command Long Range Plan,* U.S. Space Command, Peterson Air Force Base, Colorado, 1998, p. 11.

Consequently, ASATs were viewed by many at that time as destabilizing.[2] Even though ASATs were developed by both sides during the Cold War, more emphasis and investment were made in NTM and related space systems. Indeed, it was believed that the United States had a distinct technological advantage in NTM, and had more to gain in terms of intelligence collection from space because of the closed nature of Soviet society. For these reasons, it was believed that the United States had the most to lose if both sides developed increasingly sophisticated ASATs in a spiraling space arms race.[3]

Perhaps the strongest impetus for developing a U.S. ASAT capability during the Cold War derived from the fear that the Soviet Union would deploy weapons of mass destruction (WMD) in space—e.g., weapons-carrying satellites that could overfly the United States. This concern led to the Army Nike Zeus and Air Force Satellite Interceptor (SAINT) ASAT programs. Both programs were canceled soon after agreements banning the testing and deployment of WMD in space— the Nuclear Test Ban and Outer Space Treaties—were signed by the two superpowers.[4]

The ASAT debate resumed between various administrations and the Congress in the late 1960s and early 1970s after it was learned that the Soviets had tested a destructive ASAT interceptor, which led to the Air Force air-launched F-15 ASAT program. The F-15 ASAT, also a destructive interceptor system, was tested five times from 1984 to 1986. Four of the tests were reportedly successful, although independent observers have cast doubt on their validity.[5]

University researchers generally agree that the Soviets conducted 20 ASAT tests between 1968 and 1982. Later Soviet tests, in which a

---

[2]Major James Lee, *Counterspace Operations for Information Dominance*, School of Advanced Airpower Studies, Air University Press, Maxwell Air Force Base, Alabama, October 1994.

[3]Major Roger C. Hunter, *A United States Antisatellite Policy for a Multipolar World*, School of Advanced Airpower Studies, Air University Press, Maxwell Air Force Base, Alabama, October 1995.

[4]Hunter, 1995.

[5]Kenneth Luongo and Thomas Wander (eds.), *The Search for Security in Space*, Cornell University Press, Ithaca, New York, 1989.

non-radar-guidance system was reportedly tested, were failures.[6] Few if any tests have been carried out by the Soviets since that time. During the earlier stages of their ASAT program, the Soviets refused U.S. diplomatic proposals to prohibit ASAT development or use. Ironically, after many years of development, the Soviets in 1981 submitted a draft ASAT treaty to the United Nations proposing to ban the deployment of all weapons in space, including ASATs. The Reagan administration rebuffed Soviet diplomatic efforts, citing a nation's inherent right of self-defense. Indeed, this viewpoint was made a central part of administration policy and was codified in a National Security Decision Directive 42, signed by President Reagan in 1982.[7] However, the Congress began to restrict U.S. ASAT tests shortly thereafter, and eventually terminated funding for the Air Force F-15 ASAT program.[8]

In summary, the Cold War was marked by significant policy shifts on ASATs by both sides. Lack of stable ASAT policy on either side, high ASAT development costs and possible technical problems, and—perhaps most important—continued emphasis on development of NTM space capabilities, all restrained U.S. and Soviet ASAT development efforts. Space control effectively became a low-priority military mission, although various military leaders of the time displayed a remarkable level of disagreement on the importance of the space control mission.[9]

SPACE CONTROL IN THE POST–COLD WAR ERA

With the proliferation of commercial and foreign space systems, it is evident that emerging space capabilities can be exploited by potential adversaries to gain significant military advantage on the battlefield and to find and strike targets deep inside another country's territory.[10] Such systems will provide intelligence and warning data,

---

[6]Luongo and Wander, 1989.

[7]Major Anthony Russo (USAF), *The Operational Denial of Commercial Space Imagery*, U.S. Army Command and General Staff College, Fort Leavenworth, Kansas, 1996.

[8]Luongo and Wander, 1989.

[9]Hunter, 1995; Lee, 1994.

[10]*Stability Implications of Open-Market Availability of Space-Based Sensor and Navigation Information*, SAIC, McLean, Virginia, November 1995; The Nonproliferation

real time C2, and targeting capabilities to potential adversaries who previously did not have the know-how or resources to develop their own military space capabilities. In addition, these space capabilities could significantly increase the lethality of enemy long-range weapons by providing accurate targeting and weapons guidance information.

On the other hand, just as during the Cold War, the availability of high-resolution imagery of neighboring countries could enhance regional stability in certain situations by reassuring regional powers as to the capabilities and intentions of their opponents or rivals, and could prevent miscalculations by decisionmakers, thus preventing regional conflict or conflict escalation.[11]

Whether emerging space systems will be stabilizing or destabilizing influences—whether they will enhance regional stability or improve the battlefield targeting and C2 of regional aggressors—the two are not mutually exclusive. A country could and likely would gain advantages in peace as well as in war.

During the Cold War, the ASAT debate was framed within the context of the strategic conflict between the United States and the Soviet Union. Today, this debate is framed within the current defense planning context: the possibility that U.S. forces may be engaged in two nearly simultaneous MRCs. The important role space systems will play for U.S. forces during an MRC is no longer a matter of debate. U.S. military and commercial space systems provided vital support to U.S. forces during Desert Storm. Indeed, some have called the Gulf War the first "space war." The increased emphasis on space in the U.S. military has led to new military concepts of operation and new weapon systems, such as precision-guided munitions, that increasingly rely on space. Thus, in the post–Cold War era, not only may adversaries rely on emerging commercial and foreign space capabilities to gain an advantage on the battlefield, they may seek to deny U.S. military forces access to U.S. government

---

Policy Education Center, *Commercial Satellite Imagery Proliferation: A Problem to Control?* February 1995.

[11]SAIC, 1995; The Nonproliferation Policy Education Center, 1995.

or commercial satellites, i.e., they may decide to develop their own ASAT capabilities.[12]

DEFENSIVE SPACE CONTROL MEASURES

Defensive space control includes operational and policy measures. Operational measures consist of active and passive actions to protect space-related capabilities from adversary attack or interference. Policy measures deal primarily with treaties to outlaw space-based weapons of mass destruction. Passive actions are taken to reduce the vulnerabilities and increase the survivability of U.S. space systems and the services they provide. Specific measures may include the encryption of information transmitted to and from space systems, employment of redundant systems and service pathways, increased satellite maneuverability, and modifications to increase satellite survivability.

Satellite Survivability Enhancements

Several types of modifications could enhance the survivability of satellites. An enhanced integral propulsion system would add greater satellite maneuverability, so the satellite could avoid an ASAT interceptor attack. Antijam features designed into Comsats would allow the satellites to operate in the presence of jamming. Satellite shielding would protect against the effects of explosive blasts, and hardened electronics would prevent circuit upsets, short circuits, or permanent circuit damage caused by the detonation of nearby high-power microwave or nuclear weapons. However, these modifications would add significantly to the total weight and power requirements of the satellite.

Given the competitive nature of the commercial satellite market, it is generally not cost-effective for commercial satellite manufacturers to add such modifications to enhance system survivability. If they were voluntary, these types of defensive space control measures or countermeasures would not be implemented. Furthermore, DoD

---

[12]Sean Naylor, "U.S. Army War Game Reveals Satellite Vulnerability," *Defense News*, March 10–16, 1997.

would likely encounter significant resistance from industry if it tried to make such countermeasures mandatory, because such additions would decrease the competitiveness of U.S. firms in the international satellite market.

On the other hand, such countermeasures have occasionally been added to government satellites. When done for military communications satellites, procurement costs have been significantly larger than the costs for procurement of Comsats with similar communications capacity.

Policy Measures

As recounted above, at different times during the Cold War both the United States and the Soviet Union proposed to restrict or outlaw the development, testing, and use of ASATs. None of these diplomatic efforts succeeded, except to ban ASATs that employ WMD in space. While it is possible that additional policy measures could be pursued to ban all types of ASATs, this would represent a major shift in U.S. policy (e.g., space debris concerns could conceivably provide the motivation to restrict or ban debris-creating ASATs). Current U.S. space policy does not rule out the use of either defensive or offensive space control measures.[13]

Anti-ASAT Weapons or Operations

A variety of weapons and operational concepts could be used as anti-ASAT weapons or for anti-ASAT operations. For example, an enemy ASAT could be destroyed before launch by means of air attack (although ASAT launch sites may be heavily defended and difficult to attack). ASATs could be destroyed after launch but before reaching their intended targets by interceptors or directed energy weapons. Or, if an adversary store ASATs in orbit in anticipation of their use against U.S. assets, they could be attacked in orbit by U.S. ASATs.

A possible ASAT space vehicle threat would be an advanced Scud-like missile like the Taepo Dong 1 that North Korea claims to have used

---

[13]*U.S. National Space Policy Fact Sheet*, The White House Office of Science and Technology Policy, The White House, Washington, D.C., September 1996.

to orbit a satellite. Such a system could be used as a launch vehicle to deploy a large number of debris objects in LEO. Satellites in orbits that intersect the debris cloud would be damaged or destroyed by collisions with the debris. Once the ASAT warhead had detonated and spewed debris in LEO, there may be no way to counter or reverse this type of pollution of the space environment.

The most attractive anti-ASAT weapon to deal with this type of Third World ASAT threat is an ASAT boost-phase intercept system, a system that would closely resemble a boost-phase ballistic missile defense (BMD) system.

This suggests that anti-ASAT operations might be a potentially important new application for a future Airborne Laser (ABL) Program.[14] In this new mission context, an ABL could suffer from two limitations: a limited range of effectiveness against ASAT launch vehicles, and the limited number of ABL aircraft likely to be available for anti-ASAT operations. The range is limiting because ASATs could be launched from sanctuary that is perhaps out of ABL range. If an adversary also possessed tactical ballistic missiles (TBMs) and WMD that could be delivered by TBMs, theater BMD may be given precedence over anti-ASAT operations. Further study of this option is required.

A more robust anti-ASAT capability, one without range limitations and able to protect large satellite constellations in LEO, may have to be space-based to negate threats suddenly launched from sites deep in enemy countries. Such a system would closely resemble the Brilliant Eyes and Brilliant Pebbles systems and should be capable of intercepting ASAT launch vehicles before they deploy their warheads in LEO. Another option for a global satellite protection system could instead be based on a constellation of space-based lasers that would shoot down ASAT launch vehicles in their boost phase.[15] In this case, there would be a close relationship between BMD and anti–ASAT systems, and a robust anti-ASAT system could present serious anti-ballistic missile (ABM) treaty compliance issues.

---

[14]Russo, 1996.

[15]Russo, 1996.

The high cost of such systems and ABM treaty concerns imply that development of a robust anti-ASAT weapon capability would require more than just increased emphasis on defensive space control operations. A prerequisite may well be a determined effort to develop a national ballistic missile defense system and a related decision to renegotiate key elements the ABM treaty or to abrogate the treaty entirely. Until then, it is difficult to see how robust anti-ASAT weapons systems could be developed, tested, and fielded.

OFFENSIVE SPACE CONTROL MEASURES

Offensive space control includes operational and policy measures. Operational measures are designed to destroy or neutralize an adversary's space system or the service it provides through attacks on the space, terrestrial, or link elements of space systems. Policy measures deal primarily with agreements between the U.S. government and commercial or foreign satellite owners to deny enemy access to space services.

Policy Measures

Policy measures to deny access to satellites fall into two categories: those that are legally binding and enforceable (for example, policy measures that apply to U.S.-owned systems or to multinational systems operating under prior agreed-upon conditions such as INTEL-SAT) and those that are not legally binding or not enforceable by the U.S. government. Examples of the latter type are access restrictions applied by the foreign owners of a satellite system at the request of the U.S. government or shutter-control restrictions that are voluntarily agreed to by a number of governments or system owners, such as agreements not to image or sell imagery of Israeli territory.

First, we consider legally binding policy measures. Current U.S. policy stipulates that U.S.-manufactured communications satellites sold to foreign countries or corporations cannot be equipped with encrypted telemetry, tracking, and control (TT&C) links.[16] The satel-

---

[16]TT&C links control the payloads on a satellite.

lite's TT&C links could be vulnerable to external monitoring, thus making them potentially less useful for military purposes.

PDD 23 places certain legally binding policy restrictions on U.S. high-resolution remote sensing satellites. An FCC license must be obtained that requires the satellite be equipped with encryptable downlinks using an encryption package approved by the U.S. government. Such an encryption capability can prevent eavesdroppers from electronically capturing imagery sent to ground terminals by the satellite. This encryption capability need not be used in peacetime, allowing foreign minority shareholders or affiliates of a U.S.-owned remote sensing satellite to downlink imagery directly using their own ground stations. Because such encryption capabilities cannot be exported, foreign affiliates can downlink only imagery that is not encrypted.

The President, in consultation with the DoD, the Department of State, and Department of Commerce, can direct that the downlinks of satellites covered by PDD 23 be encrypted, thereby denying direct downlink access to foreign affiliates or other foreign parties. Initiation of this type of shutter control will probably require a presidential decision that would not be taken lightly because of the financial impact it could have on U.S. and minority foreign owners of these systems. However, if this encryption capability is invoked, it will deny adversaries access to the near-real-time high-resolution imagery produced by U.S. systems. On the other hand, one has to assume that regional imagery archives of foreign affiliates and even imagery archives in the United States maintained by U.S. companies will be accessible by potential adversaries during peacetime and could be used to identify and target fixed facilities around the world.

Denying access to multinational satellite systems would be even more difficult to accomplish using policy measures alone. For example, the INTELSAT charter that governs the use of INTELSAT satellites by member nations stipulates that its services are to be used only for peaceful purposes. However, this restriction can be and is interpreted in rather broad terms. Access to INTELSAT satellites may be denied to a member nation only if a majority of member states object to the system's use and if there are clear grounds to deny service. U.S. military forces have used INTELSAT services to support a variety of overseas operations. For example, during Desert

Storm—a United Nations–sanctioned military operation—the United States and its allies used INTELSAT satellites extensively. However, so did Iraq. No attempt was made to deny Iraq access to INTELSAT by invoking the INTELSAT charter, because the restrictions could have been applied equally to the United States and its allies. If the United States were to attempt to deny INTELSAT access to an adversary who was a consortium member, such a judgment could easily backfire, leading to restrictions on U.S. military use of the system. Consequently, we must conclude that international agreements governing the use of multinational systems would probably not be effective in denying enemy access to satellites. An attempt to enforce agreements such as the INTELSAT charter would represent the use of a relatively blunt and unpredictable policy instrument.

International shutter-control agreements have also been proposed to deal with the threatening capabilities implied by the proliferation of high-resolution imaging satellites. However, it could be difficult for an international shutter-control agreement to provide a high degree of assurance that enemy access had been denied to a foreign-owned satellite system. In the past, when the number of highly capable remote sensing systems was relatively small and when most such systems were owned by other advanced western countries, there was a reasonable chance that, when a U.S. request to deny satellite access was made to the satellite's owners, it would be granted. These systems did not operate in a true commercial marketplace, and the owners were not under pressure to produce profits. Diplomatic initiatives also had a greater chance of succeeding because the few foreign countries that did own such systems were often U.S. allies.[17] In the future, however, both of these conditions are more than likely not to be true. For example, U.S. diplomatic efforts are less likely to be successful in persuading foreign countries such as India and China to deny external access to their remote sensing satellites.

In summary, policy measures will be effective in denying enemy near-real-time access to U.S.-owned imaging satellites; however, they will be much less effective in doing so in the case of multinational or foreign-owned systems, or in denying access to

---

[17]For example U.S. diplomats were able to convince France to deny SPOT imagery to Iraq during the Gulf War (see Russo, 1996).

commercial satellite imagery archives that will be available in the United States or overseas.

Destructive ASATs

Destructive ASATs are weapon systems designed to permanently disable or destroy a satellite system, including ground stations that control or receive information from the satellite.

Ordinary weapon systems can be used to attack satellite ground stations. However, as mentioned earlier, ground terminals for future Comsats and some remote sensing satellites will increasingly be mobile and, in the case of Comsat terminals, small enough to be concealed and moved quickly to easily avoid direct attack. Not all satellite terminals will be as difficult to attack as an Iridium handheld telephone. However, for an increasing number of advanced satellite systems, the ground segment will no longer be a lucrative and easy target for physical attack.

The satellite itself could be a more lucrative future target. There are a number of ways of attacking and destroying satellites in space. The satellite could be destroyed by a hit-to-kill interceptor—a so-called kinetic kill vehicle. An example of such a weapon system is the Army ASAT system that has been under development.[18] Theoretically, the satellite could also be destroyed by a high-power microwave weapon or high-power laser.[19] We mention here only how these weapons would be employed, what policy concerns they may engender, and what restrictions may be placed on their development, testing, or use because of these concerns.

Destructive ASATs were controversial during the Cold War, and restrictions were intermittently placed on their development. U.S. leaders were reluctant to sanction the development of destructive ASATs because of their possible destabilizing effect on the strategic balance between the United States and the Soviet Union. Today this strategic balance, while still an important element of U.S. national

---

[18]Broad, William J., "In Era of Satellites, Army Plots Ways to Destroy Them," *New York Times*, March 4, 1997, p. C1.

[19]Russo, 1996.

security policy, is not the predominant concern in the current space control debate, because the most-threatening space capabilities that could be exploited by adversaries may not be Russian-owned.

However, other concerns and complications have arisen in the post– Cold War era regarding the development of destructive ASATs. Attacking and destroying satellites in space could have a number of significant drawbacks. First, if satellites were attacked using kinetic kill vehicles, a significant amount of debris could result from the collision or satellite intercept.[20] The growing space debris population is of concern to commercial satellite developers and operators, satellite insurers, and foreign governments. The space debris population is growing most rapidly in the orbits that are most heavily used to launch and station satellites, including orbits used by remote sensing and communications satellites in LEO. A French satellite was recently damaged by space debris and the Space Shuttle has had to execute evasive maneuvers on four missions. Destructive kinetic kill ASATs, even if designed to minimize debris, will have to impact their targets at high velocities and will inevitably create some debris and, in some cases, perhaps a great deal of it.

Furthermore, an unprecedented number of commercial and multinational satellite launches are planned to deploy satellites in LEO and GEO. Although satellite manufacturers and operators are taking steps to minimize debris, the amount of debris will inevitably grow, a prospect that concerns NASA and satellite insurers. Consequently, any type of ASAT that creates debris will be viewed negatively in the commercial and civil space sectors.

In addition to commercial industry concerns about the space debris problem, one has also to consider the possible threat of collateral damage to U.S. civil and NTM space assets. The probability of a satellite suffering a space debris impact is proportional to the total surface area of the satellite. The space system with the largest surface area to be deployed in LEO will be the International Space Sta-

---

[20]The Army kinetic kill vehicle is designed to minimize space debris by using a "fly swatter" apparatus to hit the target and disable it. However, it is not clear how well this apparatus will work, whether it will minimize debris for all types of satellite targets, and whether it will minimize debris for all possible intercept trajectories. Actual testing against real targets may be required to determine the full space debris implications of this weapons system.

tion. The next-largest class of systems in LEO, after the Space Shuttle, may be NTM satellites, and certainly the Hubble space telescope. Consequently, these civil and NTM systems may suffer the greatest damage as the space debris population grows. Because there may be a significant risk of collateral damage to U.S. civil and NTM systems if destructive ASATs were to be used in space, and as the LEO space environment becomes even more crowded with satellites, the constraints that may be placed on employing destructive ASATs could significantly reduce their operational utility and effectiveness.

The second concern regarding destructive ASATs is that they would probably be employed against only a very limited set of targets—satellites wholly owned by military adversaries of the United States. If the adversary were North Korea, Iran, or Iraq, for example, there may be no targets in this category at all. In addition, such regional adversaries could make substantial use of multinational systems like INTELSAT, Iridium, or of satellites owned by foreign powers that would otherwise remain neutral in the conflict, such as China or India. For example, it would be difficult to justify destroying a Chinese satellite if the United States were involved in a conflict with Iran in the Persian Gulf. An attack on a Chinese satellite could broaden and escalate such a conflict in way that would be detrimental to larger U.S. interests. And if the United States were to use destructive ASATs to attack a satellite owned by a multinational consortium, it could well lead to sanctions against the United States and preclude future use of the consortium's assets by U.S. military forces or even by U.S. commercial interests. If the U.S. military were precluded from using INTELSAT satellites, it could severely hamper U.S. forces because of their increasing reliance on Comsats, particularly Comsats with appropriately encrypted TT&C links. The vast majority of Comsats that now provide coverage of potential overseas conflict areas and have encrypted TT&C links are INTELSAT satellites. Future commercial or multinational Comsats that have these characteristics, such as Iridium, Globalstar, or Spaceways, will be valuable assets that U.S. forces will want to use. In addition, these assets will be partly owned and operated by U.S. investors and aerospace companies. It is therefore unlikely that the NCA would permit destructive ASATs to be employed against these large and valuable constellations of multinational satellites.

The third reason destructive ASATs may not be appropriate weapons in the post–Cold War era arises from constraints and prohibitions of arms control treaties, the ABM treaty, and the Outer Space treaty. The START agreements include prohibitions against either party interfering with NTM and early warning systems of the other party.[21] For the United States, this prohibition applies only to Russian NTM and early warning systems and not to those of other countries, although there have been discussions about extending the U.S. prohibition to NTM and early warning systems operated by countries that were once republics in the former Soviet Union. In any case, U.S. destructive ASATs could not be used to destroy Russian NTM or early warning satellites.

The ABM treaty also may preclude development, testing, or employment of destructive ASATs based on so-called exotic technologies. Examples of exotic technologies for destructive ASATs include high-power ground- or spaced-based lasers. However, these exotic technologies could also be used to destroy strategic ballistic missiles or their warheads and thus may fall under Article 5 of the ABM treaty.

Finally, the United Nations Outer Space treaty forbids interference by outside parties to satellites owned and operated by the treaty's signatories. While the definition of interference is subject to interpretation, certainly the destruction of a satellite by a destructive ASAT would constitute a clear and severe form of interference. Thus, the use of destructive ASATs would not be permitted by this treaty in peacetime. In wartime, enemy satellites could be attacked in the name of self-defense, of course, just as any other potential military asset could be. However, satellites owned by third parties that were exploited by an enemy during wartime fall into a grey area of international law. One could argue that the principle of proportional self-defense would apply in the case where such a satellite was used by an enemy to harm or destroy U.S. forces and that this would justify destruction of the satellite by ASAT attack.

In summary, destructive ASATs may not be appropriate instruments of war or of deterrence in the post–Cold War era for three reasons. First, growing space debris concerns and the dramatic increase in

---

[21]Dana J. Johnson, "The Impact of International Law and Treaty Obligations on United States Military Activities in Space," *High Technology Law Journal*, Vol. 33, 1987.

commercial activity in space suggest that U.S. civil and commercial space sectors and foreign allies will be opposed to the development of destructive ASATs, as could be the intelligence community because of the possibility of collateral damage to NTM systems. Second, destructive ASATs would probably be employed against only a small target set—the large number of highly capable multinational systems now under development would likely be off limits to destructive ASAT attack for foreign policy reasons and because of the significant financial blow such attacks could inflict on the U.S. aerospace industry. Third, destructive ASATs would probably not be employed against Russian NTM or early warning satellites, nor against satellites owned by multinational consortiums in peacetime, because of treaty prohibitions.

Nondestructive Space Control Systems

The reservations regarding destructive ASATs identified above led RAND to examine nondestructive space control concepts—systems that would prevent the functioning of the target satellite or its payloads without damaging the target. Such systems would ideally have effects that were

• only temporary,

• fully reversible, and

• localized or limited to a particular region on the earth's surface.

These types of nondestructive systems are attractive for the same reasons that destructive ASATs were found to be unattractive weapon options for the post–Cold War era. Namely, nondestructive systems could be specifically designed not to create space debris. Because they would not damage the target satellite and would prevent its functioning over only a limited region, such nondestructive systems could be used against multinational, commercial, or third-party satellites and so cause minimal damage or disruption to international satellite markets. Thus, the target set for nondestructive space control systems could be much larger than that for destructive ASATs, because the NCA would likely have fewer reservations regarding their employment. And finally, depending upon on the non-destructive system design and its associated employment concept, such systems might not raise some of the arms control and policy concerns identified above for destructive ASATs.

Although some nondestructive ASAT system concepts could be subject to treaty and policy concerns, their severity would probably be much reduced because the interference nondestructive systems would inflict on the target satellite would be much less than that caused by destructive ASATs.

For example, a number of potential nondestructive space control concepts would position the system nearby the target satellite, where it could carry out operations in the proximity of the target. It would not necessarily degrade or interfere with any internal operation or functionality of the target satellite. However, the nondestructive space control system would occupy nearby space and prevent the acquisition of communications signals or reflected energy and light from the area where military operations were being conducted.[22] The legal meaning of such "proximity operations" is today considered ambiguous by some experts. However, if analogy is made to existing maritime law, where it is legal to observe ships on the high seas at close range, board them in search of contraband, and prevent them from reaching embargoed ports, the proximity operations of nondestructive space control systems may also be considered legal. Thus, nondestructive space control systems that enforce imagery embargoes or communications blackouts of specific areas could be construed as carrying out legal self-defense activities, especially if the additional principle of proportional self-defense is applied in this case.[23]

RAND examined a range of potential nondestructive space control system options. Nondestructive space control options were assessed according to the following criteria:

- Technical feasibility

---

[22]Russo (1996) identified a number of nondestructive space system denial concepts, such as the use of low-power lasers to blind imaging satellites or electromagnetic wave transmitters to jam satellite communications uplinks or downlinks.

[23]For details on the Outer Space treaty and the principle of self-defense, see U.S. Air Force Air Command and Staff College, *Space Handbook (AU-18)*, Air University Press, Maxwell Air Force Base, Alabama, 1985.

- Assured effectiveness or vulnerability to countermeasures

- Limited regional effects

- Precise time control

- Relative system cost/size.

The system concepts with the above properties that appear to be the most promising are systems designed to work against specific types of satellite targets—Comsats or imaging satellites, but not both. For example, the nondestructive space control system designed to be a highly effective counter to imaging satellites would not be effective against Comsats if it operates only against the imaging sensor.

In terms of priorities, which type of nondestructive space control system should be developed first, a Comsat denial system or an imaging satellite denial system? First consider Comsat denial systems. Large numbers of Comsats may be available to adversaries in future conflicts. Although it would be relatively easy to deny access to these systems using nondestructive means, it may not be easy to identify which systems they are using and where their communication terminals are located. For this approach to succeed, extensive and precise intelligence preparation of the battlefield would be needed. In addition, because of the projected future growth of international Comsats, the number of systems that may have to be countered in the future could be significantly higher than the number available to potential adversaries today. Developing space control measures to counter them may not be cost-effective in all cases.[24] On the other hand, if access to Comsats could be denied to an adversary, and if other measures were taken to destroy enemy terrestrial communications networks, then much more than space-derived information could be denied to the enemy. If a comprehensive communications blackout could be imposed, it would enable U.S. commanders to carry out a highly effective counter

---

[24]An example of such a case would be an adversary that owned just a few Comsats and leased little or no Comsat capacity in peacetime. If the United States denied access to the enemy-owned Comsats, the adversary could potentially lease replacement capacity in the commercial market. However, in this case the adversary would have to pay premium prices for replacement Comsat capacity and only small amounts of capacity may be available in the spot market for that particular region. Thus, a U.S. Comsat denial capability could be effective in this case.

command and control campaign. It must be emphasized, however, that a robust intelligence capability and much preparation would be needed to successfully carry out such a campaign.

In contrast, as indicated in Chapter One, the number of highly capable imaging satellites that an adversary could access will likely be limited in the next five to ten years. Therefore, it may be a better use of limited resources to develop space control systems specifically designed to deny access to imaging satellites. Thus, the feasibility of nondestructive imaging satellite denial capabilities should be vigorously investigated.

Chapter Four

CHANGING SPACE SURVEILLANCE NEEDS

Space surveillance—the ability to detect, identify, track, and predict the position of space objects—is an essential element of space control. Space surveillance is a required ingredient for providing situation awareness of the space environment, identifying friendly and hostile space systems, and predicting when potentially hostile space systems will overfly an area of operations or interest.

SPACE SURVEILLANCE DURING THE COLD WAR

The United States has maintained space surveillance since the earliest days of the Cold War. During that time, the highest-priority space surveillance mission was to provide early warning of Soviet ballistic missile attack. Consequently, the majority of the sensors developed for space surveillance were early warning radars, such as the Ballistic Missile Early Warning System (BMEWS) and PAVE PAWS radars. These radars were optimally positioned to detect Soviet missiles launched over the north pole toward the United States, but were not designed or optimally positioned for general space surveillance. This limitation was not critical during the Cold War, because there were relatively few satellites in orbit and their positions did not have to be precisely known.

In addition, orbit prediction algorithms were developed to predict satellite positions days in advance, thus alleviating the need for complete and continuous monitoring of the space environment.

Because of the criticality of early warning data of Soviet missile attack to the NCA decisionmaking process, U.S. space surveillance capabilities were integrated into a highly survivable and reliable command and control system located at the Cheyenne Mountain Complex in Colorado. A centralized computer-processing architecture was established at the Cheyenne Mountain Operations Center (CMOC), where early warning data from DSP, ground-based radars, and other sources were processed. The CMOC system was designed to be highly fault tolerant and reliable and to provide an integrated real-time situation assessment and awareness capability for commanders of U.S. strategic forces. From the Cold War to today, the CMOC provides real-time situation awareness of the space environment.

THE U.S. SPACE SURVEILLANCE NETWORK

A key component of this early warning system is the U.S. space surveillance network (SSN), which includes ground-based radars and optical telescopes. The key product of the SSN is a predicted real-time picture of the space environment, including the identities of space objects, using near-real-time sensor data and the predicted locations of space objects derived from CMOC orbit-prediction algorithms. The SSN is composed of sensors, communications links, and elements of the CMOC processing center.

SSN sensors are shown in Figure 5. Most SSN radars are located on the U.S. border, looking outward to provide warning of a missile attack. As indicated in the figure, the number of SSN radars has been reduced since the end of the Cold War as a cost-saving measure. A number of SSN electro-optical telescopes are located near the equator to provide coverage of satellites in geostationary orbit.[1] It should be noted that while SSN radars provide an all-weather space surveillance capability, telescopes do not. In addition, telescopes frequently require cueing by other sensors for LEO objects because of their narrow field of view and the limited viewing time available for objects in LEO. Also, objects imaged by telescopes must be in sunlight to be detected, which further limits observation opportunities by these systems. On the other hand, telescopes are less costly to field and operate than high-power long-range radars.[2]

---

[1]The telescopes belong to the Air Force Space Command Ground-Based Electro-Optical Deep Space Surveillance (GEODSS) System.

[2]*Space Debris: An Independent Assessment*, National Research Council, 1996.

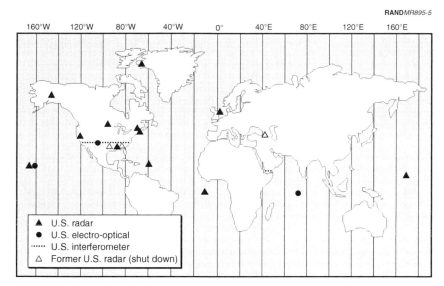

RAND*MR895-5*

▲ U.S. radar
● U.S. electro-optical
······ U.S. interferometer
△ Former U.S. radar (shut down)

SOURCE: *Space Debris: An Independent Assessment*, the National Research Council, Washington, D.C., 1996.

Figure 5—The U.S. Space Surveillance Network

As is apparent from Figure 5, the United States currently has no SSN sensors in the southern hemisphere. Because of the positioning of SSN sensors and their limited range and detection capabilities, the SSN has instantaneous coverage gaps and not all space objects can be detected directly at any instant in time. Consequently, accurate orbit-prediction algorithms are a necessary element in producing an accurate real-time space surveillance picture. These algorithms allow past observations of space objects to be combined in a systematic way with current observations to produce a complete deconflicted space surveillance picture that is valid for a specific time period if the positions of all objects in the space picture, or space catalog, are known to a minimum position location accuracy.

FUTURE SPACE SURVEILLANCE NEEDS

Space surveillance may grow in importance because of three factors: (1) the scale and complexity of commercial activities in space, resulting in many more satellites in orbit; (2) NASA concerns over space debris[3] and (3) the importance of space in U.S. and adversary military operations, and the implications for increased emphasis on U.S. space control operations. Concerns about the changing space environment reflected in the first two factors above are shared by commercial and civil space users, while the third factor is entirely military in origin.[4] In addition, the types of space surveillance capabilities suggested by these factors can differ, potentially leading to distinct technical and organizational solutions to these emerging needs.

Because of these factors, concerns have been expressed on whether the existing SSN can adequately handle the possibly more demanding space surveillance needs of the post–Cold War era. For example, can the existing SSN support future civil and commercial space industry needs, and should a civilian agency be given responsibility for some space surveillance functions? Concerns have also been expressed on whether the SSN can support emerging space control needs and whether it is keeping pace with technology advances that could help satisfy these emerging needs. We next explore potential changes in post–Cold War space surveillance needs.

Space Debris Monitoring and Warning

Since the 1960s, space surveillance has been the exclusive mission of the U.S. Space Command. Although there are indications that other nations have an interest in developing an independent space surveillance capability, the costs of doing so make it unlikely that this will happen in the near term. For example, after a French satellite was seriously damaged in a collision with space debris, French officials expressed a desire to develop an independent space surveillance capability and to reduce their reliance on the United States. Although NASA has articulated a desire for improved capabilities in space surveillance, it has done so in terms of long-term requirements, and to date has not offered to fund improvements in the

---

[3]*NASA Safety Standard Guidelines and Assessment Procedures for Limiting Orbital Debris*, National Aeronautics and Space Administration, Office of Safety and Mission Assurance, Washington, D.C., 46, NSS 1740.14, March 1995.

[4]NASA, 1995.

current system. Today, SSN radars can detect space objects in LEO with a diameter of 10 cm or larger. NASA has stated the need to detect objects with diameters as small as 1 cm.[5] Today, a vital U.S. Space Command space surveillance mission is to provide debris collision warning to NASA when the Space Shuttle is in orbit.

It would not be a simple or inexpensive matter to upgrade all SSN radars to provide the type of debris monitoring capability advocated by NASA. Today, most SSN radars operate at C-band, and would be difficult and costly to modify C-band radars to detect 1 cm diameter objects. X-band radars are much better suited for this purpose. Consequently, institution of a 1 cm space surveillance requirement may necessitate the development of new X-band radars in place of current capabilities, which would again be costly for the DoD to field and maintain. Furthermore, the fielding of multiple accurate long-range X-band radars could be viewed as a violation of the ABM treaty.

On the other hand, X-band radars designed for debris monitoring could provide a ballistic missile mid-course tracking capability that would be useful in a National Missile Defense system.

Coordination of Satellite Operations

With the hundreds of additional satellites that will be launched in the next five years and perhaps the thousands of functioning and non-functioning satellites that could be in orbit in the next decade or so, can the U.S. SSN provide an accurate and timely space picture that will meet the future needs of military, civil, and commercial users? We can speculate on whether a need will arise to deconflict satellite and launch operations of commercial and foreign enterprises, much like the FAA provides air traffic control today to competing airlines in U.S. airspace.[6] There is no civilian space surveillance counterpart to the FAA, and complicating things further is the intrinsically international nature of many satellite orbits. Today civil and military satellites of different nations share similar orbit paths with virtually no

---

[5]NASA, 1995.

[6]The U.S. Space Command provides launch windows for satellites launched from sites within the United States.

near-real-time deconfliction or coordination.[7]   As the number of satellites increases, especially in LEO, it may become necessary for some organization to coordinate satellite operations for systems owned by the DoD, by U.S. and foreign firms, and by foreign countries.

Today the United States is the dominant space-faring nation in the world, and it will likely have the most-pressing need for the international coordination of satellite operations in the coming decades, for it may have the most to lose if satellite collisions or interference were to occur frequently.

Space Surveillance and Space Control

Space surveillance is intimately connected with space control, just as air surveillance is a prerequisite for achieving and maintaining air superiority. An accurate picture of the space environment is needed to assess threats to the United States and to deployed U.S. forces (e.g., imaging opportunities of U.S. force locations by enemy surveillance satellites). An accurate space order of battle must be developed and maintained to provide situation awareness for the NCA and regional commanders. Today this need is satisfied by the U.S. Space Command Space Catalog. Satellite coordinates or orbital element sets need not be determined to great accuracy if only general situation awareness is needed, which has traditionally been the case.

Intelligence Requirements. Another and perhaps more important aspect of space situation awareness is the provision of accurate satellite identity and capability information, i.e., the accurate assessment of threat capabilities. For example, does a particular satellite have a hidden military payload? What are the payload's capabilities? Amplifying "space track" information is vital to effective prosecution of space control operations, regardless of the type of space control operation envisioned. We point out that such

---

[7]Orbital slots for GEO Comsats are coordinated before launch by the International Telecommunications Union (ITU). The ITU has taken steps to extend its Comsat orbit access authority to LEO orbits by granting frequency assignments to Teledesic and other LEO COMSAT firms. However, the ITU only confers orbit and frequency access and has no space surveillance capability of its own to monitor such agreements.

amplifying data would have to be supplied by intelligence sources and not by the SSN, although the intelligence data would have to be combined and fused with SSN data to provide an accurate space picture or Space Catalog to prevent misidentification of threat satellites. As the number of satellites and satellite developers and owners increases, threat identification may become more challenging from an intelligence standpoint.

In addition, existing space surveillance capabilities may have to be improved significantly if offensive space control operations are to be supported effectively. Depending upon the type of space control system used, the target coordinates supplied may have to be much more accurate than those typically supplied today in routine SSN operations.

Space Surveillance Needs for Nondestructive Systems. As alluded to earlier, a certain class of nondestructive space control systems appears to be preferred from the standpoint of cost and effectiveness. This class of nondestructive systems is based on a small lightweight autonomous rendezvous vehicle (ARV) that would employ a small infrared sensor to guide the system to the target. The ARV would not require inflight updates to rendezvous with the target. However, to ensure that the ARV does rendezvous with the target and not another satellite or a piece of debris, it must be equipped with a long-range high-resolution target acquisition sensor. Just as in the case of the destructive ASAT interceptor, if the initial TLE is too large, then the ARV would have to be equipped with a sophisticated target discrimination capability to distinguish the true target from background clutter and false targets. Similarly, if the initial TLE is too large, a radar-based target acquisition and tracking sensor would be needed, which would significantly increase the size, weight, and cost of the overall system. On the other hand, smaller TLEs make it possible to use an infrared target acquisition sensor, which could mean a smaller, lighter-weight, and potentially less costly ARV system.

RAND research indicates that SSN TLE accuracy would probably have to be improved by a factor of two to support this class of nondestructive space control system concepts. Since the most cost-effective and operationally useful space control systems appear to be nondestructive ARV-based systems, they should be considered as a primary input in setting future space surveillance requirements.

Space Surveillance in the Post–Cold War Era. Future space surveillance will receive increased emphasis by the military, civil, and commercial space communities. Depending upon which community's needs are given the highest priority and whether civil or commercial space users press for a nonmilitary organizational solution, emerging space surveillance requirements and system options could differ significantly.

If emphasis is given to coordinating international satellite operations, then providing accurate predicted satellite locations far in advance (perhaps weeks) of potential satellite collisions could be the space surveillance mission that is given highest priority. A space surveillance system tailored to satisfy this need would include the most-accurate satellite orbit prediction algorithms possible and the large databases needed to maintain an expanding highly accurate Space Catalog. This type of coordination activity would involve military as well as civil and commercial satellites and would benefit not just U.S. firms or the DoD. Thus, it would seem to present a case for burden-sharing, especially of sensors, with other developed nations that make extensive use of space.

If emphasis is given to monitoring the growing space debris population, including small debris objects, improved ground-based and perhaps space-based sensors would be needed. In addition, because the small debris object population may be an order of magnitude larger than what is listed in the current Space Catalog, a large state-of-the-art database engine and warehouse would be needed as well. Again, this is a mission area where the case could be made for greater burden-sharing among developed nations, since international civil and commercial users would benefit from a global space debris monitoring and collision warning system.

Finally, if the military space control mission were given highest priority, space surveillance needs would depend upon the type of space control systems that were acquired. However, regardless of the type of systems chosen, the size of the Space Catalog would remain roughly the size of the current one. If nondestructive space control systems based on the ARV concept are developed (the preferred option identified earlier), current SSN accuracy would probably have to be increased by a significant factor. Not all objects in the Space Catalog would have to be tracked at a higher-level accuracy—only those satellites deemed to be potential threats. Consequently, in this case a high-accuracy satellite track prediction system would be needed. The Air Force Scientific Advisory Board (SAB) recently studied space surveillance needs and current CMOC satellite track prediction capabilities and limitations.

> The space catalog data processing of today [in use at the CMOC] is committed to the filter and computer technologies available 40 years ago. The more accurate and efficient methods used in the commercial world and elsewhere in the National Programs community have not been exploited in the current space surveillance system. . . . The present space surveillance data processing system is tied to the missile warning data processing system in a manner that prohibits innovative solutions to the evolving space surveillance data processing problem.[8]

The SAB recommended that Air Force Space Command begin a process to modernize the hardware and software used for space surveillance data processing, improve calibration of SSN sensors, complete the upgrade of the Ground-based Electro-Optical Deep Space Surveillance (GEODSS) system, and fill gaps in GEODSS coverage.[9] There are several options for improving the satellite track prediction capabilities of the CMOC. Alternative satellite orbit prediction algorithms that could be incorporated into the CMOC and potentially provide high-accuracy satellite position predictions have been proposed by Air Force Research Laboratory (AFRL).

There are several options for improving the satellite orbit prediction capabilities of the CMOC. Satellite orbit prediction algorithms fall into three categories: analytic, semi-analytic, and numerical algorithms. The algorithm currently used in the CMOC for standard space catalog maintenance is General Perturbations 4 (GP4), which is an analytic algorithm. Semi-analytic and numerical algorithms are generally more accurate than analytic algorithms, but they are computationally more intensive and require faster computers if results are to be obtained in a timely manner. Of the three, numerical algorithms require the most computational resources, but

---

[8]*Space Surveillance, Asteroids and Comets, and Space Debris.* Volume I, *Space Surveillance*, U.S. Air Force Scientific Advisory Board, SAB-TR-96-04, June 1997.

[9]SAB, 1997.

are considered the most accurate. All three types of algorithms can be used for initial orbit determination (using observations to produce an orbital element set), differential correction (updating an orbital element set using new observation data), and orbit propagation or prediction (predicting a satellite's location based on past observations).[10]

AFRL has proposed a semi-analytic algorithm for incorporation into the CMOC. It has a program to develop PC-compatible computer code for initial orbit determination, differential correction, and orbit prediction. For this purpose, AFRL has adapted the M.I.T. Draper Laboratory's Draper Semi-analytic Satellite Theory (DSST) to run on Pentium-class personal computers. In tests, the AFRL-adapted DSST algorithms have produced timely results that are substantially more accurate than those obtained from GP4.[11] These results are encouraging and suggest that it may be possible to upgrade the CMOC to meet emerging space surveillance needs.

However, Air Force Space Command believes that a more accurate numerical algorithm—Special Perturbations (SP)—can now be implemented in the CMOC because of increases in computer processing speeds. In addition, Air Force Space Command has stated that SP is compatible with existing CMOC operations, which would make the transition to SP easier to carry out.

Regardless of which technical approach is taken in modernizing the CMOC, such modernization will likely be essential if the Air Force is to meet emerging space surveillance needs.

---

[10]Major Ted Warnock, Astrodynamic Division, Information Memorandum, Phillips Laboratory (AFMC), Kirkland Air Force Base, New Mexico, January 25, 1995.

[11]Daniel J. Fonte, Jr., *PC Based Orbit Determination*, AIAA Paper 94-3776, 1994, and Daniel J. Fonte, Jr., *Evaluation of Orbit Propagators for the Hi-Class Program*, Phillips Laboratory, Kirkland Air Force Base, PL 94-1017, 1994.

REFERENCES

Broad, William J., "In Era of Satellites, Army Plots Ways to Destroy Them," *New York Times*, March 4, 1997, p. C1.

Communications Mix Study Progress Report, MITRE, Reston, Virginia, February, 1997.

"Entrepreneurs Fashion Lockheed Martin's Strategies," *Aviation Week and Space Technology*, March 31, 1997, p 58.

"Europe Testing Technologies to Keep Pace With U.S.," *Aviation Week and Space Technology*, March 31, 1997, p. 51.

Fonte, Daniel J., Jr., *Evaluation of Orbit Propagators for the Hi-Class Program*, PL 94-107, Phillips Laboratory, Kirkland Air Force Base, New Mexico, 1994.

Fonte, Daniel J., Jr., *PC Based Orbit Determination*, AIAA Paper 94-3776, Phillips Laboratory, Kirkland Air Force Base, New Mexico, 1994.

Gabbard, C. B., K. O'Connell, G. S. Park, and P.J.E. Stan, *Emerging Markets of the Information Age: A Case Study in Remote Sensing Data and Technology*, Center for Information Revolution Analysis, RAND, DB-176-CIRA, January 1996.

Hulkower, Neal, *Update on the "Big LEOs,"* MITRE, Bedford, Massachusetts, March 7, 1995.

Hunter, Major Roger C., *A United States Antisatellite Policy for a Multipolar World*, School of Advanced Airpower Studies, Air University Press, Maxwell Air Force Base, Alabama, October 1995.

Iridium Today, Iridium Inc., Washington, D.C., Fall 1994.

"Israel Wants Imagery Ban," *Space News*, June 17–23, 1996, p. 1.

"Japan plans $1.4 billion Spysat program to counter N. Korea," *Aerospace Daily*, November 1998.

Joe, Leland, and Daniel Gonzales, *Command, Control, Communications, and Intelligence Support of Air Operations During Operation Desert Storm*, RAND, N-3610/4, 1993.

Johnson, Dana J., "The Impact of International Law and Treaty Obligations on United States Military Activities in Space," *High Technology Law Journal*, Vol. 33, 1987.

Karpiscak III, John, "Proliferation of Commercial Space Systems: Benefits and Concerns for U.S. Combat Operations," paper presented at AIAA 1998 Defense and Civil Space Programs Conference, October 1998.

Lee, Major James, *Computer Operations for Information Dominance*, School of Advanced Airpower Studies, Air University Press, Maxwell Air Force Base, Alabama, October 1994.

Luongo, Kenneth, and Thomas Wander (eds.), *The Search for Security in Space*, Cornell University Press, Ithaca, New York, 1989.

Marshall, M. F., J. Neff, N. Lao, P. Yuhas, "Military Applications of Future Commercial Space System," paper presented at AIAA 1998 Defense and Civil Space Programs Conference, October 1998.

Naylor, Sean, "U.S. Army War Game Reveals Satellite Vulnerability," *Defense News*, March 10–16, 1997.

NASA Safety Standard Guidelines and Assessment Procedures for Limiting Orbital Debris, National Aeronautics and Space Administration, Office of Safety and Mission Assurance, Washington, D.C., NSS 1740.14, March 1995.

"New Satellite Images for Sale," *International Security*, Vol. 20, No. 1, Summer 1995, pp. 94, 125.

The Nonproliferation Policy Education Center, *Commercial Satellite Imagery Proliferation: A Problem to Control?* McLean, Virginia, February 1995.

Pace, Scott, Gerald Frost, Irving Lachow, David Frelinger, Donna Fossum, Donald K. Wassem, and Monica Pinto, *The Global Positioning System—Assessing National Policies*, Critical Technologies Institute, RAND, MR-614-OSTP, 1995.

Proceedings of the Land Satellite Information in the Next Decade Conference, American Society of Photogrammetry and Remote Sensing, Vienna, Virginia, September 1995.

Proliferation of Satellite Imaging Capabilities: Developments and Implications, Berner Lanphier and Associates, Inc., February 27, 1995.

Ropelewski, Robert, "Satellite Services Soar," *Aerospace America*, November 1996.

Russo, Major Anthony (USAF), *Operational Denial of Commercial Space Imagery*, U.S. Army Command and General Staff College, Fort Leavenworth, Kansas, 1996.

Space Debris: An Independent Assessment, the National Research Council, Washington, D.C., 1996.

Stability Implications of Open-Market Availability of Space-Based Sensor and Navigation Information, SAIC, McLean, Virginia, November 9, 1995.

Technical Description of ORBCOMM Communication System and Spacecraft, ORBCOMM, Herndon, Virginia, December 1994.

USAF Scientific Advisory Board, *Space Surveillance, Asteroids and Comets, and Space Debris*, Volume 1, *Space Surveillance*, SAB-TR-96-04, June 1997.

U.S. Air Force Air Command and Staff College, *Space Handbook (AU-18)*, Air University Press, Maxwell Air Force Base, Alabama, 1985.

U.S. National Space Policy Fact Sheet, The White House Office of Science and Technology Policy, The White House, Washington, D.C., September 1996.

U.S. Space Command, *Long Range Plan,* March, 1998.

Warnock, Major Ted, "Astrodynamic Division, Information Memorandum," Phillips Laboratory (AFMC), Kirkland Air Force Base, New Mexico, January 25, 1995.